"The planetary ecological crisis begins with how we think and that makes education central to efforts to build a sustainable and decent civilization. Liza Ireland's remarkable book proposes systemic and deep changes across the board to reshape educational institutions to meet the global challenges ahead. It should be read by teachers, administrators, trustees, students, and public officials who need thoughtful answers to vexing problems."

(**David W. Orr**, *Professor, Arizona State University*)

"As an educator in the Industrial schooling system for over 20 years, I have witnessed first hand the impact such a system has on student learning and teacher resiliency, and the urgency by which change is needed. *Ecological Principles for Sustainable Education: Challenging Root Metaphors and the Industrial Schooling System* speaks to the barriers encountered by educators including myself in advocating for transformational changes, and provides management and administration with the understandings needed to bring about systemic change. Using real world examples, Dr. Ireland provides us with theory-to-practice ways to initiate the educational transformation required to transition society towards an eco-centric, life-affirming way of being."

(**Dayna Margetts**, *High School Teacher, MSc*)

"Brilliant and timely! Dr. Ireland's depth of knowledge of living systems and passion for sustainable education offers vital and experienced insights into what is possible when new epistemes recognize emergent properties and practices and we (re)solve the ways we teach and learn to see the whole of education as nested and interconnected. Our graduate students working within tired, outdated factory models of education will surely find inspiration and relief here. To imagine the eyes of the future looking back at us now, it becomes clear that the success of the student requires a profound shift in understanding of how true education is interdependent and concomitant with the health of lands and creatures, patterns and processes, and all people, present and future. This book shows us it is possible to flourish together when we live our learning."

(**Dr. Hilary Leighton**, *Program Head for the Masters of Environmental Education and Communication, School of Environment and Sustainability, Royal Roads University*)

"We finally have the blueprint for the education system that is needed for the world we are facing. Spoiler alert: What would nature do? This book not only gives the vision but explains how to implement a new system and shares real-world examples. This book ought to be read by stakeholders at every level of our current education system."

(**Erich Meyer**, *Secondary Teacher*)

"*Ecological Principles for Sustainable Education* recognizes that there is a need for opportunities and challenges to be addressed by multiple stakeholders including teacher training programs, policymakers, and elements of the general community. Case examples are provided where emphasis is placed on describing and providing adequate training and resources to teachers while promoting students with outdoor and hands-on learning experiences that can foster a deeper connection with nature and an enhanced understanding of ecological principles.

In sum, this book offers approaches that will support and enable educators to re-think and challenge many of the metaphors and practices that have shaped school systems while also offering inspiring case examples of innovative practices that have moved in new directions for sustainable environmental education. This book should be helpful to School Boards and related organizations as they assess their current situations and needs for action and resources."

(**Milt McClaren**, *Professor Emeritus, Simon Fraser University, Teacher Education*)

"*Ecological Principles for Sustainable Education: Challenging Root Metaphors and the Industrial Schooling System* is a necessary read for any person involved with educational systems. Most educational change literature focuses on one area of transformation (usually pedagogy), and merely calls for transformation. This book is not that. This book examines the educational system from all leverage points of systemic change, recognizing that the worldview of the stakeholders is of paramount importance, and calls for all educational professionals and stakeholders to examine the "why" of the system they work in, and how their own thinking and understanding of the world and themselves creates the system. It also calls for collaboration beyond a sense of what has previously been touted as "best practice" to create meaningful change, and delves into the practical how of redesigning what school is and how students are educated at every level of the system. Dr. Ireland succinctly elucidates why and how the industrial model of education sabotages previous change models and provides meaningful illumination to a path forward, out of the Anthropocene, into the Symbiocene that reflects how our planetary systems work. This book is a guide to how we need to change education, literally from the grounds up to the policy, procedure, and ways of knowing and interacting with each other and the world around us. I urge everyone to read this, share it with your friends, local school professionals and board members."

(**Nicol Suhr**, *Principal, MA Environmental Education and Communication*)

"Every now and then a book comes along that frames complex issues with clarity and resonance while providing insights to guide our way forward. Liza Ireland's *Ecological Principles for Sustainable Education: Challenging Root Metaphors and the Industrial Schooling System* is such a book. Through analysis and story Ireland leads the reader to better understand how our current approach to schooling is not fit for the purpose of building an eco-centric foundation to support a sustainable future for all living beings on the planet. The author reveals the Industrial provenance of mainstream schooling and demonstrates how a re-design for ecological principles can transform curriculum, teaching and learning, and school governance including the physical classrooms, buildings, and grounds typical for the places we call schools today. *Ecological Principles for Sustainable Education* is a fascinating and clear-eyed look at where we have been, where we are, and how we can shift course guided by the inspirational stories of innovative educators and school communities who are pointing a hopeful way forward for us all."

(**Dr. Patrick Howard**, *Dean of the School of Education, Teacher Education Standing Committee of EECOM*)

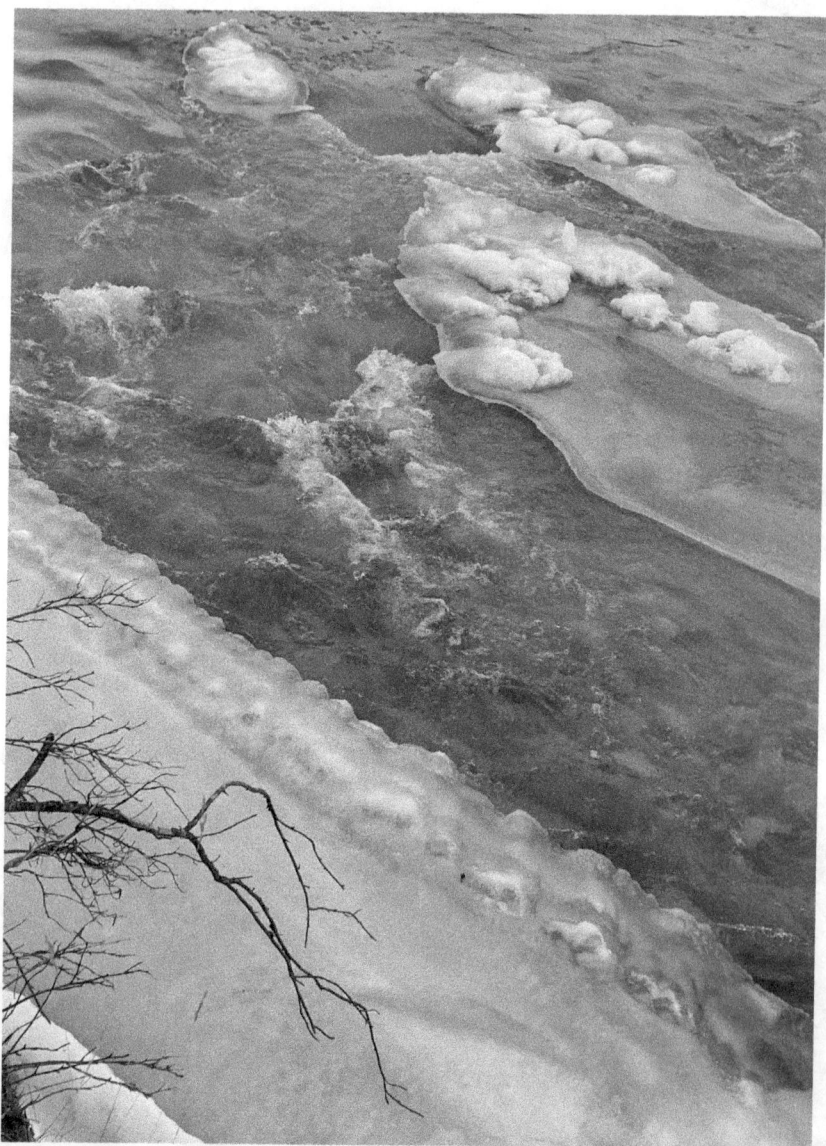

Breaking the Ice to Flow with Nature by Michelle Goulet, arts-based research in fulfilment of her MA in Environmental Education and Communication, Royal Roads University, BC.

Ecological Principles for Sustainable Education

This book explores how the education sector can transition to being truly sustainable and why necessary innovations for educational change are being subverted and undermined when mapped onto the existing industrial educational system.

Based on PhD case study research with schools that are modelling and teaching sustainability, action research, and the author's 40 years of working in the K–12 system, this volume examines how education continues to perpetuate the status quo, and why education innovations are thus undermined. It shows the importance of redesigning education based on the principles of sustainable living systems and explores how this can be achieved across all levels of the educational system. The first part of the book establishes a new vision of sustainable education, whilst the second brings to light the industrial mechanistic root metaphors in current practice across leadership and administration, buildings and grounds, curriculum design, teaching, and learning that are subverting innovative efforts. From understanding the foundational, influential, problematic root metaphors of our "Industrial" educational system, it moves to explore how the ecological principles of sustainability can be used to rethink and redesign an educational system, from its administration, leadership, and policy, to curriculum, buildings, grounds and resources, through to teaching and learning, that will support sustainability, innovation, and creativity, developing systems thinking and sustainability as a frame of mind.

Exploring how the education sector can transition to being truly sustainable and find new ways to traverse the problematic "Industrial" world view at this pivotal moment, will appeal to administrators, post-secondary educators, policymakers, and researchers and scholars of sustainability education, educational leadership, curriculum design, and educational philosophy.

Dr. Liza Ireland is Associate Faculty in the School of Environment and Sustainability at Royal Roads University, Canada, and founder of Changing Climates Educational Society.

Routledge Research in Educational Leadership series

Books in this series:

Populism and Educational Leadership, Administration and Policy
International Perspectives
Edited by Peter Milley and Eugenie A. Samier

Empirical Understanding of School Leaders' Ethical Judgements
Applications of the Ethical Perspectives Instrument
Ori Eyal and Izhak Berkovich

Culturally Sensitive Research Methods for Educational Administration and Leadership
Edited by Eugenie A. Samier and Eman S. ElKaleh

The Role of Leaders in Educational Decision-Making
Examining Implementation Factors and Providing a Newfound Model
Nancy H. Matthews

Educational Leadership and Asian Culture
Implications for Culturally Sensitive Leadership Practice
Edited by Peng Liu and Lei Mee Thien

Leading in Multicultural Schools
Cultural Intelligence and Leadership Styles for Better Organisations
Joseph Malaluan Velarde

Ecological Principles for Sustainable Education

Challenging Root Metaphors and the Industrial Schooling System

Liza Ireland

Routledge
Taylor & Francis Group

NEW YORK AND LONDON

First published 2024
by Routledge
605 Third Avenue, New York, NY 10158

and by Routledge
4 Park Square, Milton Park, Abingdon, Oxon, OX14 4RN

Routledge is an imprint of the Taylor & Francis Group, an informa business

ISBN: 978-1-032-48467-9 (hbk)
ISBN: 978-1-032-48547-8 (pbk)
ISBN: 978-1-003-38959-0 (ebk)

DOI: 10.4324/9781003389590

This book is dedicated to my grandchildren and those students, parents, teachers, administrators, and government officials willing to take the first collaborative steps in transitioning to sustainable education.

Contents

Foreword

Making a World, Making a Path

Today we have the opportunity of our lifetimes to create the future we desire. Here is the choice. We either learn to live within the limits of our planetary home and assure a liveable future for current and future generations – or we don't. The next decade or so will indicate which pathway we collectively are taking. We will boldly chart the course towards a future that is sustainable, just, and life-affirming, or we will continue with "business as usual" and thereby ensure our eventual demise.

Not so long ago, this kind of perspective was seen as alarmist and hyperbole. But now, with increasing signs of systemic breakdown across socio-economic and ecological systems documented almost daily by the news – there is an increasingly widespread sense, supported by expert opinion, that we cannot continue without fundamental change if we are to secure the future. The report to the Club of Rome *Earth for All – a survival guide for humanity,* commissioned a survey across G20 countries which revealed that 58% of people are very or extremely worried about the state of the planet, whilst 74% believe that economic priorities should "move beyond profit and increasing wealth and focus more on human wellbeing and ecological protection" (Dixson-*Declève* et al., 2022, p. 170).

Globally, we are in a state of "polycrisis" a term used by the World Economic Forum (WEF) (2023) to describe the interpenetrating knot of systemic problems that characterize our times. Global heating, instability in earth systems, loss of biodiversity, pollution, gross inequity, migration, food insecurity, overconsumption, war and conflict, the rise of fundamentalism and nationalism, and so on – it is not possible to pick out such issues in isolation from each other without regard to the complex links and dynamics involved. So how should we respond to these conditions of unsustainability, instability, and complexity?

Like a number of high-level analyses, *Earth for All* argues that there are just two broad directions or options before humanity. One is "more of the same" but with incremental policy change – this is labelled "Too Little, Too Late." The second is the "Giant Leap" and rests on the assumption that societies, spurred by recognition of the interlinked crises, quickly change course through extraordinary policy changes which are essential to build resilience and a safer future.

The difference between these two trajectories and the chance of either prevailing lie in the way we think. For the past few centuries, Western and Westernized cultures have been based on a particular way of seeing the world – a cultural worldview or paradigm – that has been enormously successful in making the modern industrialized world. This view – anthropocentric and technocentric, reductionist, dualistic, objectivist, materialist – rests on the root metaphor of "world as machine." But coupled with economic beliefs in endless material growth and expansion, individualism, and competition, the human, economic, and ecological costs of this limited view of reality are becoming increasingly apparent.

This paradigm – based upon separation and dissociation – gives rise to relationships which are frictional. We talk in dualisms – people and Nature; people and environment; economics and ecology; us and them, etc. – as if these binaries are necessarily unrelated, distinct, and in opposition. Moreover, this dis-integrated view tends to be dis-integrative and fragmentary in its effect on natural and human systems at both global and local levels. With the economy undermining the integrity of the biosphere, and heading towards climate breakdown, we are in the midst of what Joanna Macy – amongst others – refers to as the "Great Unravelling." Without fundamental transformations, we become locked into the "Too Little, Too Late" scenario.

But the alternative is clear and known. Rather than breakdown, it offers breakthrough to the Giant Leap, or (in Macy's terms) the Great Turning. It is an integrative and complementary view of people and Nature, of economy and ecology, and of present and future, and it is based on an emerging worldview which supersedes mechanism. This is the ecological – or relational – worldview, otherwise known as the "living systems" view of the world. Rather than separation, this view sees a world primarily of wholes, patterns, flows, dynamics, emergence, and complexity. It is systemic and connective and sees phenomena and events in a relational context.

It is this living systems view – so essential for regenerating our social, economic, and ecological systems particularly at local level – and so essential for regenerating hope, which inspires and informs Liza Ireland's book. For so many years, educational theory, policy, and practice have largely stayed within the parameters of mechanistic thinking, as evident, for example, by its emphasis on the individual, fragmentation of the curriculum, transmissive pedagogy, valuing the head above nurturing the heart and hands, avoidance of ethical issues, and with its purpose – particularly in the last two decades – geared primarily to boosting economic performance rather than well-being for all.

But this is changing. In particular, UNESCO has for some time recognized that educational systems are maladapted to the conditions we now face and has advocated a new and transformed vision for education. At the time of writing, UNESCO is leading a debate on "Forging a new social contract for education if humanity is to change course towards just, inclusive, and sustainable futures" (2023, p. 2). The desirability of such a future is surely beyond dispute: the question is how to make it happen – particularly here, now, on the ground, in our communities.

This is about making a world – the one we want to see and live in. The one that we owe young people, who, at a time of eco-anxiety, overwhelmingly want to make a difference that will secure their future. Awareness amongst the public, amongst parents, teachers, and particularly amongst the young is high. Understanding how to move forward with hope and with heart is less clear.

This is where Liza Ireland's book is invaluable. For many years, adding bits of environmental or sustainability education to policy and practice has been the common response of educational institutions to the sustainability agenda. Sometimes, good work has been achieved, but nearly always it was on the margins, the work of enthusiasts, whilst the mainstream remained unchanged. That's no longer good enough, if it ever was. The difference now is twofold. First, we are in a state of planetary emergency. Second, in the living systems view, we have a framework that can facilitate the whole system change that is often deemed necessary but rarely realized. This view of educational redesign is persuasive and practicable. It offers an informed critique of the mechanistic model and provides a cogent and coherent alternative which makes sense: it is constructive, positive, and life-affirming for those involved and for Nature. It is essentially regenerative of people, community, living systems – and of education and learning.

The book advocates "sustainable education" rather than "sustainability education." They sound similar, but in fact the distinction is important. The latter is a practice that is often easily accommodated and marginalized by the mainstream which stays unaffected. Sustainable education however denotes a shift of educational culture. Some years ago, I suggested that sustainable education implied four qualities that apply to educational ethos, policy, and practice, which include the following:

Sustaining – help sustain people, communities, and natural systems
Tenable – are ethically defensible, working with integrity, justice, respect, and inclusiveness
Healthy – constitute a viable system, embodying and nurturing healthy relationships and emergence at different system levels
Durable – keep evolving through continuous critical reflexivity and renewal.

Making this work at school level – bringing together theory and practice – is at the heart of this very welcome and timely book. Society is entering a rocky period like no other that has gone before. It is daunting, it is exciting. With this book, schools have a guide that will not so much light the emerging way ahead but rather help school communities envision, explore, and make the pathway to a better future.

Stephen Sterling, Emeritus Professor of Sustainability Education, University of Plymouth, author, *Learning and sustainability in dangerous times,* Agenda Publishing.

https://www.sustainableeducation.co.uk/

References

Dixson-*Declève*, S., Gaffney, O., Ghosh, J., Randers, J., Rockström, J., & Espen Stoknes, P. (2022). *Earth for all: A survival guide for humanity,* a report to the club of Rome. New Society Publishers.

UNESCO (2023). *Forging a new social contract for education perspectives on governance.* UNESCO.

World Economic Forum (WEF) (2023). *World Economic Forum Global Risks Report 2023.* World Economic Forum.

Preface

Liza Ireland

As this book is a response to our pressing need to look at how education can be transformed to help develop a sustainable society, it is carefully structured to exemplify and encourage systemic educational transformation. By developing a new societal and educational paradigm based on sustainability and systems thinking, it necessarily takes a holistic, systems view of education and is structured around a framework that inspires transformation. In doing so, it considers how philosophy, organization, administration and leadership, curriculum, buildings, grounds and resources, teaching & learning all necessarily influence each other in a multi-scale, interdependent system, being influenced by, and influencing, scales above and below in the system.

Given the issues we are facing with the dominant industrial paradigm we are currently immersed in, it is important to avoid looking at, and reinforcing, various parts of our educational system as if they are disconnected from each other. It would be easy, and some might see it logical, to fall into a mechanistic trap of looking at issues and then solutions related to various aspects such as organization, administration and leadership, curriculum, or teaching & learning separately: for example, organizing the book with a chapter on issues related to organization, administration, and leadership followed by a chapter on organization, administration, and leadership solutions; and then issues and solutions in sequential chapters related to curriculum; or teaching & learning. To do so would reinforce the dominant mechanistic paradigm, potentially encourage various actors in the system to only read a few chapters that seem to apply to their area at an individual level, thereby undermining a holistic view of the educational system.

A holistic consideration of how the various parts of the system influence each other is essential as all those involved in the educational system, from administrators to all support workers, teachers, community members, and students need to see how all these various aspects of the system are interdependent. One needs to see the whole and how the various parts interact, influencing and being influenced by other aspects of the system, in order to understand and make meaningful, transformational, systemic change. With the importance of systems thinking as a foundational construct, the book is specifically organized to look holistically at the dominant paradigm's influence in all aspects of education: organization, administration, & leadership; curriculum; buildings, grounds, & resources; teaching & learning through various the chapters in Part II. It then moves on to Part III to holistically consider how an eco-centric paradigm is reflected in creative solutions across all these aspects of the educational system. It is hoped that whatever aspect of education the reader identifies with, through this organization based on systems thinking, each will gain a systemic perspective, seeing how all aspects of the system influence the others.

To do this effectively and encourage transformational change, I have used Backcasting as a second organizational framework. It is a process and overarching organizational framework that is positive, inspiring and enables transformative change (Robinson, 2003). Backcasting starts by establishing the vision and criteria of where we need to be, before it "backcasts" to the current obstacles we face, our baseline, and then with vision and insight, moves on to look at creative solutions based on the ecological principles of sustainable living systems, and, finally, setting priorities to realize that vision (Robinson, 2003). Too often, when change is needed, we start with the status quo, with the negative problems and mindset that created the problems to begin with, getting caught in compromises and trade-offs, tacking on what seem to be various solutions to various issues in education, without a clear holistic vision of how these parts interact, or what we need to aspire to, with established criteria to weigh our solutions against. From visioning in Part I to considering the baseline of the influential dominant paradigm in all aspects of education in Part II to then developing creative solutions, based on the ecological principles of sustainable living systems, across the educational system in Part III and then setting priorities for transitioning in Part IV this book provides a holistic systems perspective to inspire, provide insight, develop creative solutions, and encourage transitional steps to achieve our vision of education for the 21st century that supports a sustainable society. This structure is in keeping with the United Nations Economic Commission for Europe (UNECE) (2012, p. 13) that identified the essential characteristics of Education for Sustainable Development (ESD) as

a A holistic approach, which seeks integrative thinking and practice;
b Envisioning change, which explores alternative futures, learns from the past, and inspires engagement in the present; and
c Achieving transformation, which serves to change in the way people learn and in the systems that support learning.

Throughout the book, I use a variety of terms related to education and sustainability. The terms Education for Sustainability (EfS), Education for

Sustainable Development (ESD), and Sustainability Education are used as these terms are common throughout the literature. However, I intentionally emphasize and use "Sustainable Education", a term introduced by Stephen Sterling (2001) to emphasize the need for systemic, educational transformation. Sterling (2008) writes,

> The concept of "sustainable education" is not just a simple "add-on" of sustainability concepts to the curriculum, but a cultural shift in the way we see education and learning, based on a more ecological or relational view of the world. Rather than a piecemeal, bolt-on response which leaves the mainstream otherwise untouched, it implies systemic change in thinking and practice, informed by what can be called more ecological thinking and values – essentially a new paradigm emerging around the poles of holism, systemic thinking, sustainability, and complexity. (p. 65)

This is an important distinction, aligning with the purpose of this book, and as such, I use this term to necessarily emphasize the transformative focus that often gets lost in much of the sustainability education discourse.

References

Robinson, J. (2003). Future subjunctive: Backcasting as social learning. *Futures*, *35*(8), 839–856. https://doi.org/10.1016/S0016-3287(03)00039-9

Sterling, S. (2001). *Sustainable education: Re-visioning learning and change*. Green Books.

Sterling, S. (2008, Spring). Sustainable education – Towards a deep learning response to unsustainability. *Policy and Practice: A Development Education Review*, *6*, 63–68.

United Nations Economic Commission for Europe (UNECE). (2012). *Learning for the future: Competencies in education for sustainable development*. Retrieved from https://unece.org/DAM/env/esd/ESD_Publications/Competences_Publication. pdf

Acknowledgements

I would like to acknowledge and thank my children Alysia, Nicholas, and Annie, and the amazing community of students and their families who trusted and collaborated with me to bring The Green School to fruition in starting this learning journey back in 1996; as well as all those students, parents, teachers, administrators, and community members in my case study schools who made this research possible.

I also want to thank my graduate students at Royal Roads University and Cape Breton University, who have agreed to share their research on how the ecological principles are helping them transition their educational practices. And a special thanks to my fellow educator Nicol Suhr for her insights in reviewing the final draft and offering examples of challenges she has faced as an administrator trying to implement sustainable education in the K-12 system. You are all pioneers.

Finally, I offer my heartfelt thanks and love to my family as they have been central to this learning journey as they have grown and flourished, as they have read and given feedback on many early drafts, and offered their constant support and encouragement to bring this book to fruition.

Part I

Envisioning the Future

If we are to develop a sustainable society, we need a clear vision of a sustainable educational system that can release us from the unsustainable mindset and practices of the industrial schooling system. In essence, if we are not guided by a conscious vision of the future we want to see, we will be guided by our unconscious acceptance of the present we already have.

DOI: 10.4324/9781003389590-1

1 Visions of a Sustainable Future and the Role of Education

As Gregory Bateson summarized so well,

> The major problems in the world are the result of the difference between how nature works, and the way people think.
>
> (referenced in Borden, 2017)

With our interconnected, dynamically changing world, we are recognizing that our essential, collective goal is to transition to a sustainable society: one that is a socially just, equitable, thriving society, imbedded in healthy, biodiverse natural environments through effective socio-ecological relationships (Dale, 2002); it is a society that functions collaboratively as part of Nature (Dale, 2002; Gunderson & Holling, 2002). To develop such sustainable human systems, we can turn to the wisdom imbedded in sustainable living systems that are based on the ecological principles of *interdependence, community, diversity, current energy flows, cycling, feedback, adaptation and emergence* (Gunderson & Holling, 2002) to maintain resilience as conditions around them constantly change. Learning from how natural living systems have become sustainable and applying these ecological principles as a framework to guide transformative change in transitioning to sustainability makes sense as humans and our social systems are also living systems, interdependent with the ecological systems we rely on (Biggs et al., 2015; Gunderson & Holling, 2002).

To become sustainable then, we need to recognize that we affect and are affected by our socio-ecological relationships, most recently exemplified by climate change and, as a result, more frequent and severe extreme weather events. Rockström and his colleagues at the Stockholm Resilience Alliance have identified nine planetary boundaries: climate change, biosphere integrity, land-system change, freshwater change, biogeochemical flows, ocean acidification, atmospheric aerosol loading, stratospheric ozone depletion, and novel entities (which include: synthetic chemicals and substances such as microplastics, endocrine disruptors, and organic

DOI: 10.4324/9781003389590-2

pollutants; anthropogenically mobilized radioactive materials, including nuclear waste and nuclear weapons; and human modification of evolution, genetically modified organisms and other direct human interventions in evolutionary processes); and the safe operating space we need to operate within (Rockström et al., 2009). In refining their metrics from 2009, it is now clear that six of the nine boundaries have been transgressed: climate change, biosphere integrity and interference with the nitrogen and phosphorous cycles, and novel entities are in the high-risk zone; while fresh water change and land use change are in the danger zone (Richardson et al., 2023). Through this research initiative Rockström concludes, "Reconciling a respect for limits with principles of justice presents the profound challenge of imagining and creating a basis for sustainable development, i.e., good lives for all on a resilient and stable planet" (Rockström, 2015, p. 6).

Given the unpredictability of the world with the complex interactions and playing out of climate change, biodiversity loss, rising inequalities, and the rise of artificial intelligence, to name a few influential change agents, we cannot predict the future and depend on it being a linear progression resembling the past. We need to understand complexity, systems, *interdependence*, and be able to embrace change and *adaptation* based on effective *feedback* mechanisms. We need to be able to *adapt* and *emerge* as conditions around us change, developing new ways to live sustainably as part of Nature. Our educational systems need to prepare us to do this, so they too need to *adapt* to the new era we find ourselves in.

There's no doubt we need to transition from our problematic, dominant industrial paradigm that falsely conceptualizes humans as separate from, superior to, and able to control Nature (Capra, 1982; Dale, 2002; Robinson, 2010; Verhagen, 2008), to a sustainable eco-centric paradigm that is the basis of sustainable living systems, recognizing we are integrally part of Nature (Pulkki et al., 2021; Rockström, 2015). To do this we need to adapt and learn how we can redesign our human systems to effectively align with Nature, seeing Nature as our model, mentor, and measure (Benyus, 2014).

Gunderson and Holling (2002), in focusing on understanding transformation in human and natural systems, look at sustainable development through an ecological lens, summarizing, "Sustainability is the capacity to create, test and maintain adaptive capability. Development is the process of creating, testing, and maintaining opportunity. So Sustainable Development is fostering adaptive capabilities and creative opportunities" (p. 76). This aligns perfectly with education and should align with all aspects of the educational system as a learning organization (Senge et al., 2012).

In order to achieve a sustainable, eco-centric paradigm shift, numerous authors over the years, from Orr (1992, 1994), Bowers (1993, 1995),

Sterling (2001, 2010, 2019), Tilbury et al. (2002), Bonnett (2007), Verhagen (2008), Benyus (2014), Robinson (2015), Selby and Kagawa (2015), the United Nations (2015), O'Brien and Howard (2016), Sterling and Huckle (2016), Hebrides et al. (2018), Pulkki et al. (2021), to Farrell et al. (2022), have recognized that transitioning from an industrial to an eco-centric paradigm is not only necessary, but it also opens exciting new opportunities for innovation, creativity, and collaboration across all sectors of society, particularly in how we educate. Orr (1992, 1994), Robinson (2015), Sterling (2019), and the United Nations (2015) argue a sustainable society depends on education where students are given diverse opportunities, skills, and are motivated to be life-long learners; are creative innovators; and effective, meaningful contributors in transitioning to sustainability. This is a society where people of all ages are excited about and love learning, see potential, explore and develop all their diverse talents, and are optimistic about and engaged in the world around them.

With this vision, our educational systems will be places of hope and inspiration seeing the major challenges of our interconnected world as opportunities to help us transition from our industrial societal models to sustainable, *adaptable communities*, based on eco-centric, systems thinking (O'Brien, 2016; Senge et al., 2012; Sterling, 2001). Pulkki et al. (2021) argue for an ecosocial philosophy of education that "cultivates an epistemic humility grounded in our ecological embeddedness in the world" (p. 347). They argue that educating the ecological self is the prerequisite of ecosocially enlightened action. This aligns with Bonnett (2007) in emphasizing we need to develop sustainability as a frame of mind. Therefore, how we educate is central to this transition.

Through the 2004–2014 UN Decade for Education for Sustainable Development (ESD) and the subsequent Sustainable Development Goals (SDGs), the UN (2015) and UNECE (2012) recognize that redesigning education, so that it plays a fundamental role in developing a sustainable society, is key in making our visions a reality. Sterling (2001), Robinson (2015), and Sterling & Huckle (2016) argue fundamental changes to how we educate are key to transitioning from our outdated industrial society and the problems we are currently facing. Robinson and Senge (2016) talk about the need to transform education from a linear, industrial model obsessed with output and yield, which destroys the culture of learning, to an organic, natural process of learning, creating the conditions where students can thrive. Bonnett (2017) argues we need to develop authentic education that centres our engaged relationship with Nature through "a fully bodied, multisensory participation in its otherness that involves feeling as much as cognition, and receptivity to intimations of fitting and unfitting response" (p. 87).

Recognizing Limitations of Earlier Efforts

There have been many educational innovations and alternative models that have brought forward different ways of educating to address our changing needs. John Dewey (1938), in the early 20th century, recognized the limitations of our industrial educational model, proposing we develop more hands-on, place-based learning opportunities. Marie Montessori and Rudolf Steiner developed alternative approaches that looked more at the whole child, connecting educational experiences to child development (Association of Waldorf Schools of North America, 2015; Montessori, 2004). These have become popular independent educational options, and we see current efforts to try to bring some of these pedagogies into mainstream teaching, but they have not had a significant impact overall.

In trying to respond to the growing environmental crisis, from the 1970s to the present, we have seen the rise of national and provincial outdoor and environmental education as well as professional associations such as the North American Association for Environmental Education, the Canadian Network for Environmental Education and Communication, as well as provincial specialist associations (such as the Alberta Council for Environmental Educators, the British Columbia Environmental Provincial Specialist Association, and the Council of Outdoor Educators of Ontario Educators) to help inspire and support teachers to bring environmental education into their teaching.

There have been an increasing emphasis, from the 1990s, on instrumental approaches to environmental education that seek to educate in, about, and for the environment with a focus on issues and taking action; as well as transformational approaches of getting children outside, and learning through experiential, place-based learning (Jickling & Wals, 2008; Palmer, 1998; Sauvé, 2005). Issue-based environmental approaches have become very popular through the Eco-schools programmes in response to the Rio Earth Summit's Agenda 21 document, which promoted local action to solve global environmental problems. FEE Eco-schools offers opportunities for schools to link with their communities and work together to solve and prevent environmental problems at the local level (Henderson and Tilbury, 2004). Instrumental ESD is well intentioned and makes some improvements to local issues in taking action, but these deal with the symptoms, not the causes. They do not change the paradigm that is causing the issues in the first place.

These environmental education approaches typically added environmental education onto the existing curriculum with an "infusion" approach that attempts to supplement the curriculum by sprinkling environmental messages and activities throughout (Henderson and Tilbury, 2004). This approach resulted in a rather fragmented diffusion of environmental education (Van Matre, 1990). As far back as the 1990s, Smyth (1995) argued that environmental education should not be a separate educational package as it came to be structured and recognized, but rather it should seek to reform education.

Further initiatives in the 21st century have continued to advocate for change. We are now seeing the rise of "forest schools" and place-based learning to try to ground children in the natural world (Child and Nature Alliance of Canada, 2023) and address the rise of what has been called "nature-deficit disorder" (Louv, 2010). These calls for innovation have created various educational initiatives, but again, they have not made significant impacts on mainstream Kindergarten to Grade 12 (K–12) education given the undermining influence of the dominant paradigm education is based on (Goodlad, 1984; Ireland, 2007; Robinson, 2010; Sterling, 2001, 2019). Even though there have been minimal changes over the years, calls for change continue, most recently with Farrell et al.'s (2022) edited book, *Teaching in the Anthropocene*, where numerous authors outline the importance of reorienting teaching based on reconciling socio-ecological relationships, social injustice, and what it means to thrive in a sustainable society. These are important pedagogical improvements, but they need an educational paradigm that will support rather than subvert them as we have seen so far.

Are Competencies the Answer?

Parallel to these calls for adding innovative pedagogy and environmental education, there have been numerous calls in mainstream education for "back to basics" so students would be prepared to take their place in the economy (Robinson, 2015). With the significant rise of computers and the digital era we now find ourselves in, there has been a call to emphasize technology literacy and adapt pedagogy, including environmental education, to incorporate technologically mediated experiences (McClaren, 2019). But is it simply about re-emphasizing the basics and bringing technological innovations into how we teach and learn, or bringing in new content, innovative pedagogy, or as we are seeing now, 21st-century competencies to address our shifting needs?

Provinces across Canada, states in the United States, and countries around the world are looking at how education needs to transform in the 21st century. Emphasis is now focused on learning competencies rather than teacher-directed, transmissive teaching of various pieces of knowledge. According to the United Nations Economic Commission for Europe (UNECE) (2012, p. 13),

> The language of competences is widely used in educational documents, including the report to the United Nations Educational, Scientific and Cultural Organization (UNESCO) of the International Commission on Education for the Twenty-first Century and the European Union recommendation on key competences for lifelong learning.

Twenty-first-century competencies fall into three broad categories: information and communication (including media literacy), thinking and problem-solving (critical and systems thinking, creativity and curiosity), personal and social (including collaboration, adaptability, and social responsibility) (Amadio et al., 2015; OECD, n.d.). Based on a review of curriculum frameworks from around the world (OECD, n.d.), many but not necessarily all these competencies are being incorporated in curricula redesigns, with systems thinking being noticeably absent. Most curriculum guides talk about individualized inquiry learning as well as collaboration; a focus on student interests, critical thinking, problem-solving, innovation, and creativity; authentic real-world learning tasks; information technology; health and well-being, adaptability, and life-long learning; developing an entrepreneurial mindset; and becoming a globally aware, engaged, ethical citizen (Alberta Education, 2011; BC Ministry of Education, 2015a, 2015b; Manitoba Ministry of Education, 2022; OECD, n.d.; Ontario Ministry of Education, 2016).

There's no doubt new policies, skills, and competencies are essential, but is simply adding these competencies to our current educational systems enough? Although we are recognizing the value of 21st-century competencies, and are revising educational curricula and pedagogy, a review of curricular documents in Australia, Africa (Ghana, Kenya), Asia (Hong Kong (China), India, Kazakhstan, Korea, Vietnam), Europe (Czech Republic, Estonia, Ireland, Scotland, Wales, Northern. Ireland, Poland, Portugal), North America (Canada, Mexico, United States), and South America (Argentina, Chile, Brazil) shows prescribed discipline-centred knowledge outcomes remain and compete with more progressive pedagogy as they continue to be the focus of both detailed subject-specific curriculum and assessment (Alberta Education, 2011; BC Ministry of Education, 2015c; Ghanaian National Council for Curriculum and Assessment [NaCCA], 2019; Government of India, 2020; Manitoba Ministry of Education, 2022; OECD, n.d.; Ontario Ministry of Education, 2019; Republic of Kenya, 2019).

With all the competing voices for adding 21st-century competencies, adding information technology, media literacy, peace education, gender education, human rights education, democracy education, climate change, biodiversity, and environmental education, as well as calls for maintaining subject silos, focusing on STEM (Science, Technology, Engineering, and Math) and "back to basics", it can be difficult to know where we turn to guide this great transformation. We are recognizing education needs to change in the 21st century but we are trying to bolt 21st-century innovations onto an outdated, flawed, industrial educational system and paradigm that subverts rather than supports the holistic, systemic thinking needed for sustainability. In essence, we are trying to get to holism through mechanism.

It is clear we are in the midst of an educational revolution with multiple competing calls for innovation, typically coming out of and supporting instrumental perspectives in either the *economy, society,* or *environment.* Each of these innovative approaches has merit, yet they seem to oppose and compete in determining how we should be transforming education in the 21st century. These competing perspectives are representative of the dominant mechanistic paradigm where the *economy, society,* and *environment* are seen as separate, typically competing, imperatives (Dale, 2002), and as a result, a traditional educational paradigm that is based on separate subjects and siloed thinking (Orr, 1992; Robinson, 2010). Rather than seeing these various emphases or economic, social, or environmental concerns as separate concerns needing to compete for attention, sustainability is about addressing and reconciling all three imperatives and developing the systems thinking needed to effectively handle this inherent complexity (Dale, 2002).

An eco-centric paradigm is based on holism and systems thinking, and the ecological principles of *interdependence* and *diversity* necessarily recognize all three imperatives and related competencies need to be addressed through how we educate. O'Brien (2016), Sterling (2019), Robinson (2015), Selby and Kagawa (2015), and the UNECE (2012) emphasize the importance of understanding and being able to apply systems thinking, believing the competencies identified above are best cultivated through an interdisciplinary educational approach that reflects the reality of the world. Catherine O'Brien coined the phrase "sustainable happiness" to bring together sustainability and systems thinking, "Sustainable happiness is happiness that contributes to individual, community, or global well-being without exploiting other people, the environment, or future generations" (O'Brien, 2016, p. 1). O'Brien (2016), Robinson (2015), Sterling (2019), and the UNECE (2012) leave no doubt that understanding complex socio-ecological and socio-cultural interdependencies, and how we can effectively overcome these issues to develop sustainably, requires systems thinking, innovation, and creativity through transforming our educational systems. Fein (2000) and Huckle (2008) have long advocated for a socially critical approach, which seeks social and environmental justice, typically challenging oppressive political structures, looking at the systemic influences that need to be addressed in developing a sustainable society.

International Efforts

The United Nations, in identifying 2004–2014 as the UN Decade of ESD, focused attention on how education could be transformed to advance sustainability, but again with an instrumental focus. At the conclusion of the UN Decade of ESD, the Aichi-Nagoya Declaration on ESD emphasized,

the potential of ESD to empower learners to transform themselves and the society they live in by developing knowledge, skills, attitudes, competences and values required for addressing global citizenship and local contextual challenges of the present and the future, such as critical and systemic thinking, analytical problem-solving, creativity, working collaboratively and making decisions in the face of uncertainty, and understanding of the interconnectedness of global challenges and responsibilities emanating from such awareness.

(UNESCO, 2014, p. 1)

Yet the Declaration also incorporates a transformational perspective inviting UNESCO member states to: "Review the purposes and values that underpin education, assess the extent to which education policy and curricula are achieving the goals of ESD" (p. 2).

Recognizing the essential role of educators in achieving ESD, the United Nations Economic Commission for Europe (UNECE, 2012) has identified competencies in ESD for educators as necessary to enable educational transformation. UNECE's (2012) competencies are based on what educators should know, what they should be able to do, how they should live and work with others, and how they should be if they are to contribute to ESD. The competences are clustered around three essential characteristics of ESD: a holistic approach, envisioning change, and achieving transformation. The holistic approach includes three interrelated components: integrative, systems thinking; inclusivity; and dealing with complexities. Envisioning change covers competences relating to the three dimensions: learning from the past, inspiring engagement in the present, and exploring alternative futures. Most importantly, achieving transformation covers competences that operate at three levels: transformation of what it means to be an educator; transformation of pedagogy, i.e., transformative approaches to teaching and learning; and transformation of the education system as a whole (p. 16/17).

The UNECE (2012) has also made general recommendations for policymakers, to provide them with a tool to integrate ESD into relevant policy documents with a view to creating an enabling environment for the development of competences across all sectors of education. These refer to professional development in education, governing and managing of institutions, curriculum development, and monitoring and assessment. Yet even with all these transformational recommendations, recent publications still focus primarily on teaching and teacher training (Farrell et al., 2022; O'Brien, 2016) in trying to drive change, ignoring the fact that teachers are caught in a top-down, mechanistic system with limited levers of change.

As we can see from the in-depth recommendations of the UNECE, the UN decade of ESD brought forward significant optimism and focus on research and various initiatives to implement ESD throughout education, but it has had limited success in transitioning our mechanistic educational system. Huckle and Wals (2015), in reviewing the impact of the UN Decade for ESD, recognized that the dominant paradigm was a significant limiting factor, "An analysis of the literature supporting the UN Decade of Education for Sustainable Development and a sample of its key products suggests that it failed to acknowledge or challenge neoliberalism as a hegemonic force blocking transitions towards genuine sustainability" (p. 493). Sterling (2016) reinforces this analysis by summarizing,

> Most policy papers, conferences, research projects and discussions on education, whether national or international, are often blind to the sustainability crisis and context that will directly affect the lives of both this generation and of those to come and moreover, reflect unexamined 'business as usual' assumptions.
>
> (p. 211)

In an effort to further advance sustainability, in 2015, 152 countries around the world contributed to and agreed to the United Nations SDGs (UN, 2015). Education is one of 17 goals, but only target 4.7 refers to ESD, limiting the target to knowledge and skills: "By 2030, ensure that all learners acquire the knowledge and skills needed to promote sustainable development, including, among others, through education for sustainable development ...." With this minimal, instrumental recognition of the role of education for sustainability across all 17 goals, and in not recognizing the need to challenge the dominant educational paradigm, Sterling sees this as a missed opportunity. In reviewing the SDGs and the role of education, Sterling (2016) summarizes,

> The role of education is more profound and comprehensive than is recognized in the text of the SDGs as regards its potential to address their implementation. Education requires a re-invention, and re-purposing so that it can assume the responsibility these challenges require, and develop the agency that is needed for transformative progress to be made.
>
> (p. 208)

Shifting the Paradigm

With these calls to review education, given the recognition and hopes for ESD to advance a sustainable society, it is clear we need to look at our current societal paradigm, and in turn our educational paradigm to see how it needs to shift so that it will support these innovations.

This will help us create an educational system that supports these goals and the development of a sustainable society that integrates all three imperatives of the economy, society and the environment; engaging, empowering, motivating, and supporting communities, administrators, teachers, and students to be innovative and creative in developing and being part of the solutions. Many youths today are in crisis, not finding meaning in school or confident in the future given the significant economic, environmental, and social crises we are facing. Education needs to be a catalyst for hope, inspiration, and skills needed in developing the vision of the future we want to be part of (Tilbury & Wortman, 2004; UNECE, 2012).

Past educational innovations are powerful as they give us a multitude of examples to work with in moving forward, but they have not been enough. We need a holistic, eco-centric guiding framework for an educational system that supports our societies transitioning to sustainability, with regenerative socio-ecological relationships, fair and equitable societies where individuals thrive and contribute to the health of the whole, and that supports systems thinking and transformation across the entire system; one that can implement the policies and competencies identified by the UNECE (2012) in addressing the UN SDGs.

We have realized how interconnected we are with the global human and non-human communities, totally reliant on natural systems and positive social and socio-ecological relationships for survival. This can inform our educational response. Fortunately, we don't need to invent anything new; we just need to turn to the wisdom imbedded in sustainable living systems to learn how we, in turn, can adapt to live and learn sustainably. The ecological principles of sustainable living systems provide the missing framework we need to guide a coherent systemic educational response. We just need to be open to systemic transformational change. As Janine Benyus says, "We are a young species, still learning how to adapt" (Benyus, 2009).

We are now at a crucial turning point in education, so it is essential we stand back to look at the foundations and influential root metaphors of our educational systems that maintain and support the status quo, undermining many well-intentioned educational innovations and efforts to transition to a sustainable society (Dale, 2002; Robinson, 2010; Sterling, 2001). This vision of education for sustainability in the 21st century is shared by communities, educators, and ministries of education around the world as they are immersed in improving and revising their provincial, state, and national curricula (Tilbury et al., 2002), yet we are typically unaware of the problematic foundations of our traditional educational systems that remain and undermine our best intentions (Ireland, 2007; Robinson, 2010; Sterling, 2019). Given the opportunities before us, at this great time of challenge and transitioning, it is essential we take a systemic perspective to recognize the industrial foundations our educational systems are built on, so we can then look forward with both insight and direction to build a new system on a solid foundation that will support innovation and transitioning to a sustainable future.

This book will explore where we've come from, reveal how hidden metaphors of the dominant paradigm, underlying all aspects of our educational system, undermine innovations and perpetuate the unsustainable status quo, and how ecological principles of sustainable living systems can provide the foundational framework to develop educational systems and approaches that not only incorporate lessons from previous educational initiatives but also guide us to develop the ethos, policies, and competencies needed to transform to a sustainable society guided by sustainability as a frame of mind (Bonnett, 2007). In essence, this guiding framework can reconcile competing interests to guide a holistic, informed educational transformation.

This book is the outcome of my PhD research; over 40+ years as a formal educator in the K–12 system in trying to implement environmental education, systems thinking, and education for sustainability; my eventual design of The Green School; as well as research from my graduate students who are educators applying this eco-centric framework in their practices. After years of having my, and my colleagues', efforts to bring environmental and sustainability education into the formal K–12 curricula subverted, I developed The Green School in Scotland from 1996 to 2000 in response to the need to put theory into action. With significant successes realized in developing The Green School, I subsequently completed my PhD research to look at why, after 30+ years, we weren't getting anywhere in the formal system. My PhD case study research asked:

1   What are the influential conceptual metaphors involved in executing a viable curriculum for education for sustainability?
2   To what extent does the use of specific metaphors in the context of philosophy, policy formation, organization/administration, curriculum development, and teaching and learning practice within select elementary schools hinder and promote education for sustainability? and
3   What do the teachers, administrators, students, and parents perceive to be the successes, obstacles, and needs in developing models of good practice in education for sustainability?

In addressing these questions, this book considers fundamental, systemic changes to our conceptualization and practices in education.

In Part II, we will consider our traditional educational constructs in philosophy, organization/administration, curriculum, where education happens, as well as teaching and learning, supported by the case study research from a government-run elementary school that decided to model and teach sustainability as a whole school approach. With these insights we will be able to understand how our educational system has served and continues to maintain the status quo, and the importance of redesigning the foundations of education across all areas to enable effective transformative change.

Part III then looks at how we can effectively redesign education so that it supports rather than subverts sustainability by creating an educational system based on the ecological principles of sustainable living systems. This enables and supports systemic, creative solutions across all sectors of our K–12 educational systems, including teacher training. This section will be supported by case study research of an independent elementary school that was designed on bioregionalism, the development of The Green School (Scotland) that was intentionally designed based on ecological principles and outdoor experiential learning. Part IV then provides suggestions and practical examples of transitioning to help inspire transformations, from my PhD research, The Green School, graduate students as practising educators, as well as a few examples from other educational institutions at various levels in education.

It's time we get past window dressing in education, get past simply adding technology or new competencies, curricular units, or asking teachers to implement progressive pedagogy that the system itself doesn't support. It's time we get past rearranging deck chairs on the Titanic and build a new ship that can help us navigate the complexity and rapidly changing currents in the 21st century.

References

Alberta Education. (2011). *Framework for student learning: Competencies for engaged thinkers, ethical citizens with an entrepreneurial spirit.* Retrieved from https://open.alberta.ca/publications/9780778596479

Amadio, M., Opertti, R., & Tedesco, J. C. (2015). *The curriculum in debates and in educational reforms to 2030: For a curriculum agenda of the twenty-first century.* UNESCO. Retrieved from http://unesdoc.unesco.org/images/0023/002342/234220e.pdf

Association of Waldorf Schools of North America (2015). *Waldorf education.* Retrieved from https://waldorfeducation.org/waldorf_education

BC Ministry of Education. (2015a). *BC's education plan: Focus on learning.* Retrieved from http://www.bcedplan.ca

BC Ministry of Education. (2015b). *Curriculum redesign.* Retrieved from https://curriculum.gov.bc.ca/rethinking-curriculum

BC Ministry of Education. (2015c). *BC's curriculum.* Retrieved from https://curriculum.gov.bc.ca/curriculum/overview

Benyus, J. (2009). Biomimicry in action. *TED Talk.* https://www.ted.com/talks/janine_benyus_biomimicry_in_action?language=en

Biggs, R., Schluter, M., & Schoon, M. (2015). *Principles for building resilience: Sustaining ecosystem services in social-ecological systems.* Cambridge University Press.

Bonnett, M. (2007). Environmental education and the issue of nature. *Journal of Curriculum Studies, 39*(6), 707–721.

Bonnett, M. (2017). Sustainability and human being: Towards the hidden centre of authentic education. In B. Jickling & S. Sterling (Eds), *Post-sustainability and environmental education: Remaking education for the future*. Palgrave Studies in Education and the Environment.

Borden, R. J. (2017). Gregory Bateson's search for "patterns which connect" ecology and mind. *Human Ecology Review, Society for Human Ecology, 23*(2), 87–96.

Bowers, C. A. (1993). *Education, cultural myths, and the ecological crisis: Toward deep changes*. State University of New York Press.

Bowers, C. A. (1995). *Educating for an ecologically sustainable culture*. State of New York Press.

Capra, F. (1982). *The turning point. Science, society, and the rising culture*. Bantam.

Child and Nature Alliance. (2023). https://childnature.ca/

Dale, A. (2002). *At the edge: Sustainable development in the 21st century*. UBC Press.

Dewey, J. (1938). *Experience and education*. Macmillan.

Farrell, A., Skyhar, C., & Lam, M. (Eds.). (2022). *Teaching in the anthropocene*. CSP Books Inc.

Fein, J. (2000). 'Education for the environment: A critique' – An Analysis. *Environmental Education Research, 6*(2), 179–192.

Ghanaian National Council for Curriculum and Assessment (NaCCA). (2019). *Development of the Ghanaian curriculum*. Retrieved March 2023, from https://nacca.gov.gh/curriculum/

Goodlad, J. (1984). *A place called school. Prospects for the future*. McGraw-Hill Book Company.

Gunderson, L.H., & Holling, C. A. (Eds.) (2002). *Panarchy: understanding transformations in human and natural systems*. Island Press.

Government of India. (2020). *National education policy 2020*. Retrieved March 2023, from https://ncf.ncert.gov.in/assets/uploadedfiles/documents/NEP_Final_English_0.pdf

Hebrides, I., Affifi, R., Blenkinsop, S., Gelter, H., Gilbert, D., Gilbert, J., Irwin, R., Jensen, A., Jickling, B., Knowlton Cockett, P., Morse, M., Sitka-Sage, M., Sterling, S., Timmerman, N., & Welz, A. (2018). *Wild pedagogies: Touchstones for re-negotiating education and the environment in the anthropocene* (B. Jickling, S. Blenkinsop, N. Timmerman, & M. D. Sitka-Sage, Eds.). Palgrave Macmillan. https://www.palgrave.com/gp/book/9783319901756

Henderson, K, & Tilbury, D. (2004). *Whole-School Approaches to Sustainability: An International Review of Sustainable School Programs*. Report Prepared by the Australian Research Institute in Education for Sustainability (ARIES) for The Department of the Environment and Heritage, Australian Government.

Huckle, J. (2008). An analysis of new labour's policy on education for sustainable development with particular reference to socially critical approaches. *Environmental Education Research, 14*(1), 65–75. https://doi.org/10.1080/13504620701843392

Huckle, J., & Wals, A. E. J. (2015). The UN Decade of Education for Sustainable Development: Business as usual in the end, *Environmental Education Research, 21*(3), 491–505, https://doi.org/10.1080/13504622.2015.1011084

Ireland, L. (2007). *Educating for the 21st century: Advancing an ecologically sustainable society* [Doctoral dissertation, University of Stirling]. STORRE. http://hdl.handle.net/1893/240

Jickling, B., & Wals, A. E. J. (2008). Globalization and environmental education: Looking beyond sustainable development. *Journal of Curriculum Studies, 40*(1), 1–21.

Louv, R. (2010). *Last child in the woods.* Atlantic Books. Cited.

Manitoba Ministry of Education (2022). *Curriculum kindergarten to grade 12.* Retrieved from https://www.edu.gov.mb.ca/k12/cur/science/index.html

McClaren, M. (2019). Revisioning environmental literacy in the context of a global information and communications ecosphere. *The Journal of Environmental Education, 50*(4–6), 416–435.

Montessori, M. 1870-1952. (2004). *The Montessori method: The origins of an educational innovation.* Rowman & Littlefield Publishers.

O'Brien, C. (2016). *Education for sustainable happiness and well-being.* Routledge.

O'Brien, C., & Howard, P. (2016). The living school: The emergence of a transformative sustainability education paradigm. *Journal of Education for Sustainable Development, 10*(1), 115–130.

OECD. (n.d.). *National or regional curriculum frameworks and visualizations annex.* Retrieved March 2023, from https://www.oecd.org/education/2030-project/curriculum-analysis/National_or_regional_curriculum_frameworks_and_visualisations.pdf

Ontario Ministry of Education. (2016). *21st Century competencies: Foundation document for discussion.* Retrieved from http://www.edugains.ca/newsite/21stCenturyLearning/about_learning_in_ontario.html

Ontario Ministry of Education. (2019). *Education that works for you.* Retrieved from https://news.ontario.ca/en/backgrounder/51527/education-that-works-for-you-modernizing-learning

Orr, D. (1992). *Ecological literacy: Education and the transition to a postmodern world.* State University of New York Press.

Orr, D. W. (1994). *Earth in mind: On education, environment, and the human prospect.* Island Press.

Palmer, J. (1998). *Environmental education in the 21 century: Theory, practice progress and promise.* Routledge.

Pulkki, J., Varpanen, J., & Mullen, J. (2021). Ecosocial philosophy of education: Ecologizing the opinionated self. *Studies in Philosophy and Education, 40,* 347–364. https://doi.org/10.1007/s11217-020-09748-3

Republic of Kenya. (2019). *Basic education curriculum framework.* Retrieved March 2023, from https://kicd.ac.ke/wp-content/plugins/pdfjs-viewer-shortcode/pdfjs/web/viewer.php?file=https://kicd.ac.ke/wp-content/uploads/2019/08/BASIC-EDUCATION-CURRICULUM-FRAMEWORK-2019.pdf&attachment_id=&dButton=true&pButton=false&oButton=false&sButton=true#%5B%7B%22num%22%3A92%2C%22gen%22%3A0%7D%2C%7B%22name%22%3A%22XYZ%22%7D%2C69%2C228%2C0%5D

Richardson, K., Steffen W., Lucht, W., Bendtsen, J., Cornell, S. E., et al. (2023). Earth beyond six of nine Planetary Boundaries. *Science Advances 9*(37). Retrieved from https://www.science.org/doi/10.1126/sciadv.adh2458

Robinson, K. (2010). Changing education paradigms. *TED Talk.* Retrieved from https://www.ted.com/talks/sir_ken_robinson_changing_education_paradigms

Robinson, K. (2015). *Creative schools: The grassroots revolution that is transforming education*. Penguin Books.

Robinson, K., & Senge, P. (2016). *Sir Ken Robinson and Dr. Peter Senge: Education fit for the 21ˢᵗ century*. Disruptive Innovation Festival. Retrieved from https:// www.youtube.com/watch?v=j1egR1szeH4&list=PLIGwcUQqMXyguvLdrkTazf 4a8BniyVN0x&index=1&t=0s.

Rockström, J., Steffen, W., Noone, K., Persson, Å, Chapin, F. S., Lambin, E., Lenton, T. M., Scheffer, M., Folke, C., Schellnhuber, H., Nykvist, B., de Wit, C. A., Hughes, T., van der Leeuw, S., Rodhe, H., Sörlin, S., Snyder, P. K., Costanza, R., Svedin, U., … Foley, J. (2009). 'Planetary boundaries: Exploring the safe operating space for humanity'. *Ecology and Society, 14*(2), 32–65. Retrieved 1 June 2016, from http://www.ecologyandsociety.org/vol14/iss2/art32/

Rockström, J. W. (2015). *Bounding the planetary future: Why we need a great transition*. Tellus Institute. Retrieved 1 June 2016, from http://www.tellus.org/pub/ Rockstrom-Bounding_the_Planetary_Future.pdf

Sauvé, L. (2005). Currents in environmental education: Mapping a complex and evolving pedagogical field. *Canadian Journal of Environmental Education, 10*(1), 11–37.

Selby, D., & Kagawa, F. (Eds.). (2015). *Sustainability frontiers: Critical and transformative voices from the boarderlands of sustainability education*. Barbara Budrich Publishers.

Senge, P., Cambron-McCabe, N., Lucas, T., Smith, B., Dutton, J., & Kleiner, A. (2012). *Schools that learn: A fifth discipline fieldbook for educators, parents, and everyone who cares about education*. Crown Business.

Smyth, J. (1995). Environment and education: A view of a changing scene. *Environmental Education Research, 1*(1), 3–120.

Sterling, S. (2001). *Sustainable education: Re-visioning learning and change*. Green Books.

Sterling, S. (2010). Learning for resilience, or the resilient learner? Towards a necessary reconciliation in a paradigm of sustainable education. *Environmental Education Research, 16*(5–6), 511–528. https://doi.org/10.1080/13504622.2010. 505427

Sterling, S. (2016). A commentary on education and sustainable development goals. *Journal of Education for Sustainable Development, 10*(2), 208–213. https://doi. org/10.1177/0973408216661886

Sterling, S. (2019). *Becoming 'learner drivers' for the future: Re-thinking learning and education in a troubled world*. Routledge Sustainable Development Pages, May 2019.

Sterling, S., & Huckle, J. (2016). *Education for sustainability*. Routledge. (Book originally published in 1996, republished in hardback in 2016 with new chapters). https://www.routledge.com/Education-for-Sustainability/Sterling-Huckle/p/ book/9781853832567

Tilbury, D., & Wortman, D. (2004). *Engaging people in sustainability*. IUCN.

Tilbury, D., Stevenson, R. B., Fein, J. & Schreuder, D. (Eds.). (2002). *Education and sustainability: Responding to the global challenge*. IUCN. https://books. google.ca/books?hl=en&lr=&id=q18nBgAAQBAJ&oi=fnd&pg=PA13&dq=

Curriculum+innovation+creativity+sustainability&ots=Kn97G6ijZI&sig=Gihoh QYcBh48Ja_bPFgepQ9pztk&redir_esc=y#v=onepage&q&f=false

United Nations (UN). (2015). *The UN sustainable development goals.* Retrieved from https://sdgs.un.org

United Nations Economic Commission for Europe (UNECE). (2012). *Learning for the future: Competencies in education for sustainable development.* Retrieved from https://unece.org/DAM/env/esd/ESD_Publications/Competences_Publication. pdf

United Nations Educational, Scientific and Cultural, Organization (UNESCO). (2014). *Aichi-Nagoya declaration on education for sustainable development.* Retrieved 1 June 2016, from https://sustainabledevelopment.un.org/content/ documents/5859Aichi-Nagoya_Declaration_EN.pdf

Van Matre, S. (1990). *Earth education: A new beginning.* The Institute for Earth Education.

Verhagen, F. (2008). Worldviews and metaphors in the human-nature relationship: An ecolinguistic exploration through the ages. *Language and Ecology, 2*(3), 1–19.

Part II

What Is Holding Us Back?

The significant problems we face cannot be solved at the same level of thinking we were at when we created them.

(Albert Einstein)

DOI: 10.4324/9781003389590-3

2 The Story of School

Einstein's insight that a new type of thinking is essential for us to survive (Amrine, 1946) is just as applicable today. Before we can realize this vision of an educational system needed to help develop a sustainable society, we need to "backcast" to the present to identify what is holding us back and what needs to change.

After all my years in school, as a student from Kindergarten to PhD; as a teacher at all levels, trying to help so many students learn and deal with school; as a parent trying to help my own children through school; and as an academic and member of society looking at the need to develop a sustainable future, I've been asking a lot of questions ... Why isn't school working as we hope it will? Why isn't it preparing our youth to deal with the challenges of the 21st century? And why do we educate the way we do? Are kids really happy, developing to their full potential? Why do so many children start out wanting to go to school and then after three to four years often look for excuses to get out of going? Why do we often feel like cogs in a machine? Does school have something to do with our reluctance to engage effectively with our complex challenges in society, the economy, and the environment? Above all else: *Is this the best we can do?*

As we engage with the many great challenges in transitioning to a sustainable society, it's essential we look at our educational system, and the influence it has had in developing the mindset and thinking that has led to the unsustainable status quo. David Orr, as far back as 1992, summarized, "The environmental crisis is first and foremost a crisis of the mind, perception and values and hence a challenge to those institutions presuming to shape minds, perceptions and values. It is an educational challenge." (p. 27). So how did we get to where we are, living in an unsustainable world, where schools are having a hard time connecting with students and the world we are experiencing, where schools are maintaining the status quo rather than helping us transition to a sustainable society? As we saw in the introductory chapter, if we are to transition to developing a sustainable society, we need to look closely at our societal and resultant educational paradigm.

DOI: 10.4324/9781003389590-4

Essentially, we need to identify what is needed to transform education, based on what is undermining transformation. To do this, we need to understand the roots of our educational system and the problems we are experiencing. It is important to take a closer look at our dominant paradigm – how our mechanistic, industrial paradigm or worldview has become imbedded in our educational systems such that education has been serving to maintain this status quo. By looking at the root foundations and metaphors that influence how and why education happens as it does, we can then understand why numerous, insightful, well-intentioned teaching approaches have been subverted by the system itself, and why the recommendations coming out of the UN Decade for Education for Sustainable Development (ESD) have not come to fruition. This will give us the insight needed to be able to design a new system that will support necessary innovations; one capable of navigating the complex reality and challenges before us – one that carries us forward, transitioning from an outdated system that is based on a limited, outdated worldview to one that develops a sustainability mindset based on the principles of sustainable living systems.

Influences of the Dominant Societal Paradigm

In order to chart a new course, we need to step back and look at history and the foundations of modern formalized education that led us to where we are today. To understand the extent and depth of this challenge, Bowers (1993) emphasized it is necessary to look at the underlying societal paradigm that shapes our perceptions, values, and guiding (often taken-for-granted) metaphors that, in turn, shape and direct our educational efforts.

The story of modern school has been developing from the Scientific and Industrial Revolutions when linear, mechanistic thinking became extremely influential in our Western society. Roots of the dominant scientific, industrial paradigm that governs our Western society can be traced to Rene Descartes in the 17th century, as well as thinkers such as Newton, Bacon, and Locke who furthered Descartes' ideas (Capra, 1982). The significant paradigm shift that took place focused on how we can understand and control our world, and what makes humans special.

Descartes concluded that mind and matter were separate. He is credited with the famous phrase, "Cogito ergo sum," I think therefore I am, seeing the mind as separate from the body and Nature. He asserted that: "there is nothing included in the concept of body that belongs to the mind; and nothing in that of mind that belongs to the body" (quoted in Capra, 1982, p. 45). This separation of mind and matter was to become a major factor in Western thought, opening the door to a less spiritual, more objectified view of life. "I think, therefore I am" influenced by this same emphasis and separation in the educational system. The rational mind was seen as separate from and superior to the bodily, the spiritual, the emotional, and the intuitive in relationship with the natural environment (Orr, 1992).

Descartes challenged the traditional, faith-based ways of knowing and came up with a new way of seeing the world like a machine, understanding it through enlightenment reasoning. He felt that if we could take things apart to understand how they work we could then predict and control our world ourselves (Capra, 1982). For Descartes, and for Newton after him, that which is not mind is machine. The human body, animals, plants, and the natural world were seen as mechanical and as such they are perfectly understandable by reducing them to their separate component parts. Wholes were seen to be made up of separate parts; the whole being no more or less than the sum of its parts. Connections and interactions were not considered important. Once viewed as a machine the non-human world could then be utilized and manipulated to extend our control over the Earth believing this would lead to a better future (Capra, 1982).

Isaac Newton, influenced by Descartes in the age of the Scientific Revolution and Enlightenment Reasoning, furthered these mechanistic ideas in developing mathematical formulas and laws to explain the physical universe and how everything is predictable. In Newton's universe, individual bodies such as planets, followed these laws, exerting force on one another while staying fundamentally independent. This mechanistic thinking encouraged people to focus on individual parts, breaking problems down into pieces and then arranging them in a logical order to understand them and develop the ability to predict and control our world. This reductionist thinking led to a multitude of new discoveries and many medical advances, but again, connections between these parts were not considered as important as the part itself. The whole was understood as no more than a mere sum of its parts and unpredictable outcomes were seen as the result of an external force (Orr, 1992).

Through the 1800s and early 1900s, rational scientific thinking was thought to be neutral and culture free as it was based on reality, on what the scientist actually observed in the natural world. Thought to be value-free, scientific thinking was considered to be the height of intellectual achievement. It was reasoned that humans alone had superior intelligence capabilities. As such, humans were considered to be separate from, more important, and more intelligent than other species. It was believed to be humankind's right to exploit Nature for our own benefit (Schwartz & Schwartz, 1995). We have continued to emphasize enlightenment reasoning, thinking there is no connection between intuitive emotional thoughts, deductive reasoning, and logic – although cognitive science is now showing us emotions are centrally important to reasoning as the emotional part of our brain guides decision-making (Lakoff, 2010).

As a result of this industrial paradigm shift, mechanism still remains in many underlying, taken-for-granted metaphors that guide our thinking either consciously or unconsciously (Bowers, 1993; Smith, 1992). Schwartz and Schwartz (1995, p. 168) summarized the generative root metaphors of the culture of modernism to include:

> (1) anthropocentrism, (2) the belief in unending progress, (3) the belief that linear, rational thought (neutral, natural, and culture free) is the epitome of intellectual achievement, (4) the belief in the dualisms of mind/body and humankind/nature, (5) the belief in humankind's right to exploit nature, (6) the belief in market economics (both capitalist and socialist), (7) the belief in the metaphor that society functions like a machine and that humans function as individual independent units of this machine, and (8) the belief that society is best controlled when power is centralized.

It is important to recognize these mechanistic root metaphors as they have a powerful influence on education.

Educational Implications

In Europe and North America this new mechanistic thinking was helping transform society. This linear, mechanistic thinking has continued to influence and guide our social development, particularly education, right up to the present (Goodlad, 1984).

In using reductionist thinking, education was developed to operate like a machine and humans were seen to function as individual, independent units of this machine. As such, they needed to be controlled: and it was believed that *society is best controlled when power is centralized* (Schwartz & Schwartz, 1995). Napoleon was a master of centralized control, having defeated the unruly Prussians in 1806 with his well-disciplined, controlled army. Crushed by their defeat, the Prussians felt they needed to instil more discipline and have greater control over their citizens, and to bring this about they designed a government educational system that would turn out well-disciplined citizens who would follow orders without questioning authority. It was "scientific" and based on centralized control. But, interestingly, within this compulsory educational system, the elite were allowed to follow a different approach that encouraged them to think for themselves while the masses were conditioned to be compliant. Children needed to be "schooled," with the aim to make compliant citizens able to contribute to the rapidly developing society of the 19th and 20th centuries (Robinson, 2015).

The Prussian model provided an effective factory system for education based on discipline, centralized control, and a mechanistic, linear approach to learning. Knowledge was broken down into separate subjects like science, mathematics, language arts, physical activity, art, and music to be taught in different periods of the day (with the central core of reading, writing, maths and science seen as the most important parts). Each discipline was then broken down further so it could be efficiently transmitted in a linear format from kindergarten to graduation. In addition to compulsory attendance, the Prussian system demanded national training for teachers, national testing for all students (important because it gave the government the ability to classify children for potential job training), national curriculum set for each grade, and mandatory kindergarten (Hern, 2003).

In the 1900s this system of education was specifically brought to Massachusetts in the United States by Horace Mann and Edward Everett and was soon after adopted in New York and across North America. By adopting the Prussian system, education could become carefully controlled and prescribed by a central higher authority. H. L. Mencken wrote in The American Mercury for April 1924 that the aim of public education is not:

> to fill the young of the species with knowledge and awaken their intelligence. … Nothing could be further from the truth. The aim … is simply to reduce as many individuals as possible to the same safe level, to breed and train a standardized citizenry, to put down dissent and originality. That is its aim in the United States … and that is its aim everywhere else.

Although this represents a dated and negative response to a carefully controlled and prescribed system of education, the educational model is based on the efficient, economical transmission of knowledge. It does not recognize the diverse needs of the learner as a person in various contexts as it does the need to be cost-effective in producing the necessary workforce to keep the production/consumption, consumer economy moving (Robinson, 2010; Senge et al., 2012). This obviously frustrates the many teachers whose focus on learners' needs is at odds with the objectives of the system and controlling governments. With its emphasis on workers for the economy, the emphasis is on knowing that has instrumental value. Today, schools are often forced to compete for government funding based on their performance following a business model of performance goals, external evaluations, and league table comparisons. Competition is central to this model as success is judged in terms of reaching the highest levels by achieving externally imposed performance criteria and ranking national educational systems based on schools' and individuals' performance (Sjøberg, 2015).

Throughout the 20th century many have complained about students being controlled by society and not being treated as creative individuals able to have much say. The formal educational system does not have a history of supporting the whole child learning in relationship. The mechanistic foundation dismisses completely the environmental context life and learning takes place in, as well as the holistic Nature of children who learn through all their senses, and through social and socio-ecological networks (Robinson, 2015). One of the more famous critics of the traditional mechanistic, linear educational approach, Albert Einstein, had this to say:

> One had to cram all this stuff into one's mind, whether one liked it or not. This coercion had such a deterring effect that, after I had passed the final examination, I found the consideration of any scientific problems distasteful to me for an entire year.

The educational emphasis became strongly focused on the individual becoming a useful contributor to the developing industrial economy and the industrial view of society (Robinson, 2010).

It is not surprising there were numerous progressive educators advocating for a different way to educate. Rudolf Steiner challenged the industrial model in education emphasizing education needs to be built around developmental stages and engage the whole child, yet he was also influenced by mechanism, separating education from natural contexts, insisting science should never be taught outside, but exclusively in a laboratory (Association of Waldorf Schools of North America, 2015). In contrast, Dewey and Rousseau emphasized experiential education in line with the development of the whole child and education as a medium between Nature and society (Boyles-Deron, 2006; Dewey, 1938; Oelkers, 2002). From the early 1900s John Dewey has been a major contributor to holistic, democratic educational philosophy and practice, advocating students be actively engaged in the learning process through hands-on, experiential education. However, these progressive educational philosophers were also influenced by the dominant philosophy of their day as they encouraged the development of individualistic trends (Gilead, 2005) and instrumentalism (Boyles-Deron, 2006). Paulo Friere's work is similarly criticized for its anthropocentrism focusing on child-centredness and the student's immediate experience, and neglect of the ecological crisis (Bowers, 1993). For Friere, rational thought, in the form of critical reflection, is the only source of authentic, legitimate knowledge. Ultimately, although Dewey's, Rousseau's, and Friere's ideas are inspirational and make sense in how to make learning interesting, meaningful, and relevant, they have not been widely applied as they are not easily implemented in a system that is not structured to support this way of learning.

The 1960s social movements further challenged centralized control, emphasizing individual expression and freedom. Adam Curtis in "The Century of the Self" shows the 1960s saw a significant reinforcement of the individual and individual choice, in an attempt to break free of what were considered restrictive social controls (Curtis, 2002). With the changes in society in the 1960s, education also started to speak about individuals being able to express themselves, about promoting child-centred learning approaches, specialist schools, electives, and parental involvement. By the 1970s there was support for experiential/outdoor education, open classrooms without walls, and child-centred approaches. Although mainstream education made some superficial changes and appeared to focus on the individual, the government retained careful control by continuing mandatory schooling, issuing a prescribed curriculum; imposing standardized testing and graduation requirements; and requiring and controlling teacher certification. This centralized control overpowered many progressive innovations and the system never really changed (Robinson, 2015).

Numerous societal concerns have continued to challenge how we educate. The 1980s questioned the quality of education seeing a swing from left to right with the "back-to-basics" and performance standards movement arguing students needed to be competitive in the growing global economy (Goodlad, 1984). The 1990s and 2000s have continued the standards-based curriculum focused on learning outcomes and competitive league tables to compare and rank schools (Sjøberg, 2015). More recently, we are seeing calls to respond to the realities of a highly digital, networked world of information technology, shifting from learning outcomes to instrumental competencies in response to PISA rankings, as well as continued calls for the need to encourage innovation and creativity through progressive teaching methods (Robinson, 2016).

Caught in the middle, teachers are trying to implement creative, innovative, pedagogy while having to respond to the undermining, controlling influence of centralized assessment focused on separate subject disciplines. Throughout the 20th and 21st centuries educators have made valiant attempts to reform education and implement progressive pedagogy, yet innovations have become subverted or sidelined with very little lasting effect. As the educational system, and the underlying thinking it is based on, has not changed and is very powerful, innovations that run counter to the mechanistic framework tend to be subverted or are illusionary at best (Ireland, 2007; Sterling, 2001, 2016).

In every aspect of schooling our educational systems continue to maintain the status quo and entrench the outdated mechanistic worldview. Let's explore how various aspects of the educational system currently teach this outdated mindset through both its overt and hidden curricula by considering the mechanistic root metaphors: *the world works like a machine and individuals are independent units of that machine; society is best controlled through centralized, top-down control; the economy, society, and environment are separate; the world is in our hands – humans can and have the right to control the environment; nature is a resource; rational thought (neutral, natural, and culture free) is the epitome of intellectual achievement; take the world apart to understand it; progress is linear with knowledge comprised of separate linear building blocks;* and *survival of the fittest.* Integrated into the various sections below are quotes from practising teachers, who were in my 2022 curriculum design course for a Master of Education in Sustainability, Innovation, and Creativity, reflecting on how the mechanistic system is currently influencing their practice. Findings from my PhD research on how these mechanistic root metaphors influenced and ultimately undermined efforts to model and teach sustainability in a Canadian government-run elementary school, that grafted sustainability onto their taken-for-granted mechanistic educational system, and experiences of a practicing school administrator are added in the "Systemic Effects" sections.

Organizational Structure & Administration

The organizational structure and administration of our educational system is based on the root mechanistic metaphors of *society is best controlled when power is centralized* and that *the world works like a machine*, compartmentalizing various parts of the overall system in separate silos where boundaries are maintained (Sterling, 2001). Through this structure, educational change tends to happen by planning and imposing in a linear direction, from the top-down (Figure 2.1), and we manage things or people by command and control (Ireland, 2007; Senge et al., 2012).

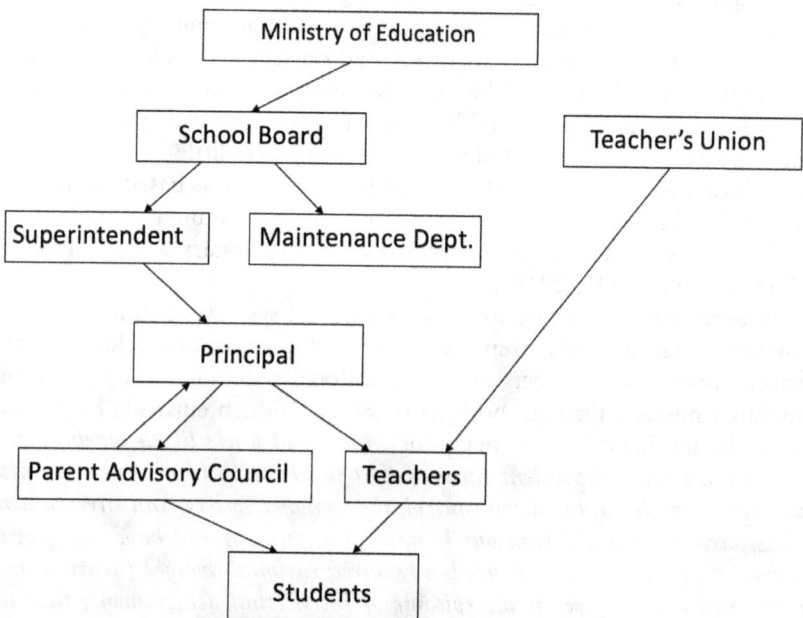

Figure 2.1 Top-down centralized control. Created by the author.

As exemplified by Figure 2.1, education has developed this very top-down, centralized control hierarchy with rigid compartments, with the Ministry of Education at the top, making decisions that are to be carried out or passed down by school boards. School boards give directives to the district superintendents, as well as maintenance departments that function independently as a separate department. Superintendents then provide directives down to principals, principals to teachers, and then teachers down to students at the bottom of the hierarchy. Parents are involved to a minor degree through meetings with Parent Advisory Councils in speaking to principals about various school initiatives that then influence students. Teachers are also influenced by their unions, which in turn affects students when a teacher's strike is called. Outside of the organizational structure, there is typically some linear communication between teachers and parents regarding student learning. Throughout the hierarchy, there is little to no feedback or influence from lower scales to those above. Students are not involved in decisions regarding their education at any level but are seen as mere recipients in this top-down process (Ireland, 2007).

Ask any student and they know they're relatively powerless on the bottom; and their role is to jump through the hoops set out for them. Even with curricular and pedagogical innovations in the 21st century, control continues to be centrally maintained through a fairly rigid top-down hierarchy, controlled by each successive level in the hierarchy (Ireland, 2007; Robinson, 2015). From the Ministry of Education come curriculum directives, approved courses, competencies, and intended learning outcomes for all subjects at all levels K–12 (Alberta Ministry of Education, 2011; BC Ministry of Education, 2015; Manitoba Ministry of Education, 2022; OECD, n.d.; Ontario Ministry of Education, 2019). These policies are directed through the school board to the District Superintendent, to School Administrators, to Teachers, and finally delivered to students. Parents and students have very little say in the process. Feedback loops are minimal at best (Ireland, 2007).

This industrial organizational structure creates issues with trust. There is less trust between the district administrators and schools as they aren't active participants in schools and they aren't as visible. As a result, when issues do arise at the school level, district administrators typically come in from a top-down, centrally controlled perspective feeling they need to "step in" and "fix things". Even when permission is granted for an innovation to go ahead, issues often arise as innovations are not supported by the industrial system schools are imbedded in. District administrators often do little to remove barriers that emerge during the process, usually because of collective agreement concerns with a union and human relations (HR) rules designed to reduce conflict between the players rather than encourage collaboration.

As a result, innovations from lower levels are difficult to realize, and trust is eroded as organizational siloes are maintained and reinforced.

Systemic Effects

The negative influence of mechanistic root metaphors in administration and leadership was exemplified in my case study research, where a government-run elementary school optimistically set out to model and teach sustainability, with buy-in from the school administration, teachers, students, parents, support staff, and wider community. To help with leadership they worked with a consultant and also sourced an outside grant to hire a Sustainability Co-ordinator from the school community.

Centralized control, silos, and the maintenance of clear boundaries were evident at all levels of the hierarchy. Financial and policy controls were very centralized in the Ministry of Education and, as the elementary school was also a community school, the Ministry of Families and Children was also involved. Changes at the ministerial levels directly affected the types of programmes that could be offered as well as available staff and finances.

When asked if they could work with the school board to support their grassroots efforts the Program Co-ordinator replied,

> I went to a meeting and had a ten-minute presentation relating the economics of sustainable actions. From that we were invited to come down and do an environmental audit for the School Board. The students did it but nothing else has happened since. I wrote a letter saying teachers need to feel empowered. The response was the Superintendent needed to balance his budget so he could not offer financial support.

Noting centralized control by the school board affected the finances of the sustainability programme, the Program Co-ordinator explained:

> When you teach and live sustainably, you save money. The problem is if our school saves the school district money in the running costs of the school, the district does not allow the school to have that money. They argue they need it as they are running at a deficit.

Overall, those involved in this grassroots initiative recognized that the financial needs of developing a programme needed to be built into it right from the start. However, the Program Co-ordinator recognized those in charge of the budget did not prioritize sustainability. As she noted, "Considering the political climate of schools right now, working in a deficit, they are going to go for the cheapest thing. It would really be incredible if they manage to do the budget and be sustainable." This showed how centralization and a lack of a shared philosophy and commitment to sustainability were seen as obstacles.

So although the school and community school were in agreement that they would model and teach sustainability, this agreement was not consistently implemented throughout all levels of the hierarchy. Their grassroots initiative did not get support or backing from other levels of administration. The Principal said, "So far it has proven it has been our own school, our own initiative and gumption to make it happen. We've had no financial support at all and very little other support from the district level."

Beyond finances, centralized staffing decisions, that the school had no control over, had the potential to seriously impact the longevity of the programme. Although the community school raised the money to pay for the Sustainability Co-ordinator, as that person would work in the school, the hiring of a Sustainability Co-ordinator was controlled by the school board and the Teacher's Union. Also, the school could not request that new teachers hired have a background in sustainability, as that is dictated by the local Teacher's Union: they could say it would be useful but being a primary trained teacher is the only prerequisite that could be insisted on, and the Principal was rarely involved in hiring decisions by the school board. Yet, as Greig et al. (1989), Orr (1992), Bowers (1993), Sterling (2001), and UNECE (2012) have argued, it is the classroom teacher who would need to implement the programme for sustainability to become central to the curriculum.

Centralized control affected the principal as well as teachers. As the Principal seemed to be the catalyst between the community school and the elementary school, he played a pivotal role in implementation and development of the programme. Unfortunately, principals are centrally controlled by the school board and are often moved to other schools, thereby affecting the leadership and direction of a school and its programmes. This is exactly what did happen. The Principal was transferred to another school in the district when the Sustainability Co-ordinator's position also ended, due to a lack of funding, and as the new Principal did not have a background in sustainability they were unable to fill the void in leadership.

This lack of systemic support and functioning in separate silos was also exemplified by the School Board Maintenance Department as it worked independently from the school, even though their decisions affected the school buildings and grounds where learning happened. This was apparent during Earth Week when the school was showcasing its sustainability programmes to the community and the Grounds Maintenance crew showed up to spray toxins to exterminate Carpenter ants. A teacher who was preparing the debut of her class video production had no idea they were coming and was very surprised by the lack of co-ordination and communication between departments and levels:

> For example, who authorized an exterminator to poison an outbreak of Carpenter ants which aren't life threatening – during Earth Week at that! I don't know who was concerned whether it was the School Board being concerned about their buildings, whether teachers complained, or it was a health issue. The timing was too stressful with setting up for the video premier and the Principal was rushing out, so I didn't say anything to the kids and I felt badly. But many times it comes down to where are you going to put your energies; and where do you have power; and where is it appropriate to exert that power.

This exemplifies the lack of communication and understanding that was common in this mechanistic organizational structure. The clear defence of boundaries, minimal decentralized control, and poor communication often caused conflicts between what the school and community school originally envisioned and what actually happened.

Further lack of support and defence of boundaries stemming from centralized control was exemplified by the teachers and their union during the second year of the sustainability programme when the teachers were involved in a union dispute with their employers. In the second year of the sustainability programme, when the Co-ordinator was trying to influence teachers to incorporate sustainability into their teaching, the Teacher's Union instructed teachers to employ work-to-rule tactics so all extra-curricular or special programmes not in the actual employment contract were contentious issues. As the Sustainability initiative was not mandated by the Ministry of Education, it was suspended. The Co-ordinator spoke about the obstacle of the teachers defending their boundaries, created by the job action, as specifically affecting their involvement in the sustainability programme. When the Co-ordinator was asked if she met regularly with teachers to give them feedback on the curriculum learning outcomes and what they could do regarding sustainability, she replied:

> No, not this year given the whole political climate. The teachers were involved in contract disputes and job action. Last year was a lot more flexible. This year it has been very difficult to have a working relationship with people who are totally stressed out by the government. There has been a lot of anger and resentment and drawing of lines. I'm also not teaching this year, so I don't have that kind of connection with the Intermediate Team. They talk amongst themselves.

Evidence of working in silos and defence of boundaries was evident even though there was wide consultation and basic agreement to model and teach sustainability.

Much of the administration and leadership issues resulted from trying to graft a sustainability programme onto a mechanistic system. The management structure depended on individuals taking control of various initiatives with top-down control, and synergies and emergence were not emphasized. When asked how sustainability initiatives were developed, a teacher replied:

> Usually we discuss it at a staff meeting, which happens once a month. People have an opportunity to say if they like it or dislike it. If it looks like a large portion of the staff are for it, usually one or two people will take it on, push the rest of the staff to do it, get meetings organized, get all the stuff done. If there are no people who spark onto it, it doesn't get done. You need to have one or two people that will be in charge of it. It can't be a whole staff initiative.

This shows that although the staff has to be accepting of an idea, the emphasis is on a few people to drive it forward to make it happen rather than through an ecological view that supports co-operation, collaboration, and shared responsibility (Sterling, 2001). The development of the sustainability programme followed this format of a few interested people initiating and being the driving force. The Principal recognized that this surface level of commitment was all that was sought:

> The Program was sold to the teachers as a program that would not add to their workload. And that was a problem as much as I think we needed to do that because again, if we said this is what we believe in, this is what we want and we want you to create everything and develop on your own, forget it.

A teacher on staff, and the Community School Board Chair, recognized the top-down control they developed, purposely taking responsibility off everyone's shoulders so they would get agreement to move the programme forward. Once approval had been received from the teaching staff, school administration, and the Parents Advisory Council (PAC), a few committed people from the community school continued to move the programme forward. In doing so, with virtually no input from others, they started to alienate others from being involved or developing a sense of ownership and commitment. The Principal noted,

> We had the support of all the staff so there were a few people who went ahead and wrote the proposal and got the funding. But the

staff didn't own that proposal; they didn't have any input into it other than they knew this was an area they wanted to get into. That was a problem.

Students were also constrained by the mechanistic management structure as student involvement often depended on teacher support. Being at the bottom of the hierarchy in a school system structured on top-down control, students did not have power to make changes unless teachers supported and authorized those actions. The Co-ordinator tried to involve the students by establishing a Student Advisory Council in the second year. When students on the Advisory Committee were asked if they felt teachers were open to their ideas they replied, "Some are really open but some don't want to listen, they don't have the time." The students felt the teachers, not the students, held the power to decide what ideas would or would not be followed up on. As one Grade 6 student, referring to the absence of recycling containers in the classroom, put it, "It is kind of their choice because it's their classroom."

Beyond the Student Advisory Committee, a lack of communication and limited opportunities for involvement seemed to curtail greater student involvement. When asked about student involvement, Grade 6 students showed their frustration with the lack of communication and empowerment by saying, "We don't know what [Student Advisory Committee] are doing really. They don't really tell us what they do. Some of us wanted to be on the Committee and [the teachers] wouldn't let us." Some Grade 7 students become so frustrated with the lack of empowerment, not being asked for their ideas and being told what to do, they reported subverting what was happening by throwing their compost in the paper recycling. In the end, the students, with the help of a parent, started a Sustainability Kids Club outside the school, so they could be involved.

By the follow-up visit, two years on, significant administrative changes threatened to negatively impact the sustainability programme. It no longer had a paid Co-ordinator as the external grant had expired. This, combined with a change in Principals, resulted in sustainability no longer being a regular item on meeting agendas, and there was no strong leadership driving the programme forward. In addition, the community school, funded through the Ministry of Families and Children, was threatened with staff layoffs and closure. As the community school initiated and managed the sustainability programme, continued support or growth within mainstream education came to an end.

This administrative approach is indicative of the dominator cultural orientation as it gives priority to top-down control, power, and authority rather than prioritizing the lessons from Nature that reveal life develops by networking not dominating (Senge et al., 2012). Another example of the ineffectiveness of trying to impose changes from the top down was experienced by Nicol Suhr, a practicing school administrator, part of a team working to enhance Indigenous learning, counter to the status quo:

> A district-wide planning group initiated a learning framework to enhance indigenous learning and movement from anthropocentric views to holism. Professional learning opportunities are provided, transportation made available, different times, types of events are facilitated, including remote opportunities with books provided, social media, technology such as podcasts, community-based events with food, music, storytelling. Yet buy-in is limited. It is viewed as "one more thing" and teachers and administrators view the initiative as exhausting, time consuming and not relevant to the work they are required to accomplish in the regular school day. Despite the inherent belief and obvious value of the initiative to the wellbeing of students, the active implementation is still regarded as too cumbersome to put into place as regular practice. The framework that led to the development of the initiative is adopted as a primary guide for decision making for the district, but few people remember it, utilize it, or understand how to communicate with it (that includes the people who actually wrote it!).

In order to challenge the status quo and redesign our unsustainable mechanistic educational system, we really need to rethink our organizational structures and administrative practices that are designed to maintain the status quo and retain centralized, top-down power through domination that does not encourage and support innovation to emerge.

Buildings/Grounds/Resources

Buildings, school grounds, and resources have traditionally been seen as external to the learning process, having been designed and supplied at minimal cost (Goodlad, 1984) yet they are powerful aspects of the hidden curriculum. As the emphasis on education has been seen as a process of gaining decontextualized, instrumental, predetermined knowledge through rational thought, school buildings were designed to provide shelter, desks, and chairs, isolated from the community and natural world. This organizational structure, determining where and how we learn, is very influential as part of what we learn through the hidden curriculum, reinforcing the root metaphors *the economy, society, and environment are separate; the world is in our hands – humans can and have the right to control the environment, nature is a resource, the world works like a machine, and individuals are independent units of that machine.*

Reinforcing the belief that the *environment, society, and economy are separate,* our school buildings tend to separate students and educators from our environment reinforcing the false notion that we learn best through the transmission of decontextualized knowledge, separate from our human communities and more than human environments.

Our school buildings and the resources used to support them are often wasteful of energy, water, and resources. Schools often produce significant amounts of waste, use toxic cleaning materials, and are not designed to be carbon neutral or minimize water use, thereby reinforcing the root metaphors *the economy, society, and environment are separate* and *the world is in our hands* such that *Nature is seen as an endless resource and waste sink* for humans, where economic considerations are more important than social or environmental imperatives.

School buildings and grounds have typically not encouraged transformative, active, experiential learning and development through all one's senses, in context with our community and environment (Robinson, 2015). As a result, playgrounds have traditionally been fenced, paved areas for playing in a human-constructed context, separate from the natural environment and wider community; while learning happens isolated in the classroom. Encouragingly, there has been a significant movement to green school grounds and develop outdoor classrooms with the support of NGOs such as Destination Conservation, Learning for Sustainable Futures, the Alberta Council for Environmental Educators, and the Child and Nature Alliance of Canada, to name a few. However, even with these developments, the mechanistic structure of the curriculum and scheduling limitations are such that teaching is still largely focused in the classroom (Goodlad, 1984; Ireland, 2007).

Systemic Effects

In the government-run school that tried to model and teach sustainability, they were very aware of the buildings and grounds needing to be part of their initiatives. However, they are owned and maintained by the local school board. The Program Co-ordinator felt there were problems with the school not having control over the grounds:

> The leaf blower came for eight hours a day. They go for what is the quickest and cheapest way to maintain the grounds. Maintenance uses leaf blowers to blow the leaves from here to there on a windy day causing excessive noise pollution, and sidewalks are power-washed to get dirt out of cracks when we are trying to reduce our water consumption by 10%. There is not a good connection between our school and the School Board who contract maintenance.

A Kindergarten/Grade 1 Teacher recognized there were many ecological changes she would like to see but were beyond her control. She was very concerned with the chemical cleansers the custodial staff used in the class, especially on the surfaces of the desks and tables, not only in concern for the students but for everyone exposed to them.

Unfortunately, the centralized control over buildings and grounds also affected joining programmes that would make retrofitting the school more energy and economically efficient. Unfortunately, the Maintenance Department would not put the money needed up front even though they would recoup it and more in two years. As a shared philosophy in sustainability did not exist between the school and the school board, the school board used short-term economics as their main consideration, thereby limiting what the school could or could not do. They did not see the important connections between the economy, society, education, and the environment. The hidden curriculum was powerfully teaching the environment and sustainability are not important considerations.

Another example of centralized control limiting adoption of a proven energy innovation was shared by a practicing school administrator, Nicol Suhr, following her master's research in environmental education and communication:

A staff member from a different district noticed that none of the schools have solar panels to assist with electricity generation. This staff member came from a school district where all the school buildings had extensive solar arrays, despite having multiple feet of snow per winter; where installation was conducted with work experience credit for students who also gained solar panel installation professional certification. When bringing this to the attention of the school principal and to district administration, they were told that solar panels were too heavy for the school buildings and it was inefficient and cost prohibitive to put them on. When questioned about the potential for working with the Canadian Union for Public Employees as the other district did, the staff member was told that this union wouldn't permit it, and the district couldn't see opening up the collective agreement to explore such an option.

Such unwillingness to explore potential innovations that are happening elsewhere in defending the status quo, stymies innovation and sends messages to those seeing the need to initiate change that sustainability is not important or a matter for education. We need to pay attention to what we are teaching our students through the hidden as well as the overt curriculum.

Curriculum

The overt curriculum is also extremely mechanistic through the root metaphors of *centralized control, taking the world apart to understand I* (reductionism), *I think therefore I am* (cognitive, analytical, left-brain emphasis), and *linear thinking.* As we saw earlier, the curriculum is prescribed and controlled by centralized governments at the top of the hierarchy. It is based on atomistic, reductionist thinking and enlightenment reasoning, which breaks knowledge into subject-disciple-focused bits and parts spread across and reassembled in the linear educational system. It is based on the belief that students will pick up each successive piece, like building blocks, progressing from simple to complex. As such, the traditional mechanistic curriculum embodies reductionist, linear thinking through detailed, decontextualized, and abstract knowledge.

Centralized Control

As demonstrated in the "Organizational Structure & Administration" section, the Ministry of Education determines the curriculum through centralized control, planning, and imposing the curriculum and any changes to it from the top-down. Although instrumental competencies are the most recent additions to curricula, these are expected to be taught through a subject discipline structure such that students are prepared to use this disciplinary knowledge in the developing economy (Alberta Ministry of Education, 2011; BC Ministry of Education, 2015; Ghanaian National Council for Curriculum and Assessment (NaCCA), 2019; Government of India, 2020; Manitoba Ministry of Education, 2022; OECD, n.d.; Ontario Ministry of Education, 2019; Republic of Kenya, 2019).

Reductionism: Taking the World Apart to Understand It

Figure 2.2 shows how a typical curriculum structure is based on reductionism, reinforcing the root metaphor *take the world apart to understand it.*

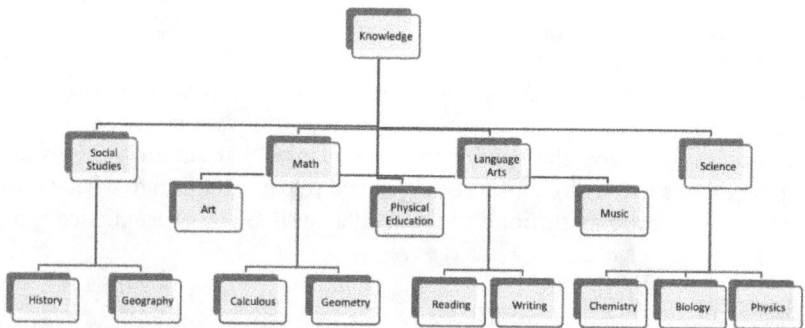

Figure 2.2 Knowledge based on separate subject disciplines.

Figure 2.2 exemplifies some of the main subject disciplines in the K–12 educational system, such as maths, language arts, science, and social studies, that are more highly valued; as well as art, physical education, and music that are seen as secondary in importance. These subject disciplines become further broken down into more discrete areas in designing curricula and learning outcomes. The further you go in education, the more segregated, specialized, and decontextualized subject disciplines become. There are many more specific categories that could have been illustrated, such as listening and speaking in language arts as well as further specializations such as poetry. Art could be further broken down into pottery, drawing, art history, or photography; music into jazz band or musical history; and physical education into various sports, dance, or health education. Even though learning competencies have made their way into recent curriculum developments, knowledge continues to dominate, separated into subject disciplines, and within each discipline, concepts and knowledge have been broken down into learning outcomes. Noel McInnis (1982) saw this as "thinking the world to pieces" (p. 85).

This curriculum model reinforces reductionist thinking through this structure, and by focusing on separate discipline subjects – as if science is unrelated and separate from mathematics or social studies; or that communicating well is only important if you are in an English class. Hence, the common student misconception is that spelling shouldn't count in social studies or science. The decontextualized curriculum teaches, for example, you cannot learn to read until you can name all the letters in the alphabet, subtract before you add, or explore three dimensions before you've studied two dimensions. A practising teacher in my 2022 curriculum design course reflected on how they taught maths in a traditional curriculum, "[You] break down math into units as well. First you learn multiplication, then division, followed by fractions and then 2-D shapes. Once you were done looking closely at one topic, you would move on to the next."

This mechanistic structure is firmly entrenched in typical day in an elementary school which starts with language arts (broken into reading and writing exercises, and spelling from a random list of decontextualized words), followed by recess, then maths, lunch, and either science or social studies in the afternoon after the "serious" learning of the 3Rs (reading, writing, and arithmetic) has occurred. Physical education as well as French as a second language is often sprinkled throughout the week with a separate specialist teacher; and art or music is typically scheduled on a Friday afternoon. High school curricula are even more siloed and subject-specific with students having to turn off their English learning to go to social studies, and then turn that off when the bell rings to move to maths, and then again turn off maths to go to science. A practising high school teacher recently reflected,

> When I think about the metaphor of "take the world apart to understand it" I think about how so much of our school day is divided up by subject areas. Teachers/curriculum makers place the different aspects of the world into different categories with the thought that by doing so, we will have a deeper understanding of it. By focusing just on math, biology, chemistry, history, reading, etc., we can look closely at the different parts of the world and learn about it more deeply in isolation.

However, the parts are rarely put together, in context. As another practising teacher elaborated,

> The job is broken down into time blocks, and we walk into the building, teach our curriculum, and leave sometimes without talking to other departments, or other adults some days. The world is taken apart so that students can earn their credits, which total to a High School diploma, but there are additive pieces that we are missing.

The student is left with a head full of disjointed bits and pieces of knowledge, collected as they progress through the educational assembly line from Kindergarten to Grade 12, and up through university.

Reinforcing Linear Thinking

In treating students like machines on an assembly line, isolated in age-specific classrooms, the factory model of educating, the top-down control, and the sense that we progress by getting from A to B suggest a linear way of thinking and acting. Everything about schools – from progression through kindergarten to Grade 12, to how curricula are designed, and how subjects are taught – suggests a linear progression and teaches linear thinking. In speaking about how knowledge is broken up into separate building blocks that need to be assembled through a linear progression of learning in a clearly defined order, a practising teacher in my 2022 curriculum design course recognized this is a common practice,

> As a Science teacher, I tend to hear many teachers discussing the importance of specific knowledge outcomes and that the sequence of these building blocks of knowledge needs to be taught in a specific order so that students in the class can learn the "logical" progression of their course themes ... I personally feel that too much time is given

in High School to students explicitly teaching content for the purpose of factual recall, so much so, that teachers don't get to the creation of new knowledge with students as they run out of time.

Another practising teacher recognized this has significant repercussions:

> Curriculum organized in a linear/building block fashion organized by age and Grade level also gives the message that learning only happens at school. At what point can we start giving measurable credit for learning that students do outside of the classroom?

This linear system and linear, mechanistic, reductionist thinking are not sufficient or effective when we are dealing with the complex problems of the 21st century and trying to transition to sustainability.

Systemic Effects

When the government-run elementary school in my PhD research decided to model and teach sustainability, it followed the British Columbia (BC) Curriculum for Elementary grades. This curriculum was defined in terms of learning outcomes for the subjects of English language, arts, mathematics, sciences, social studies, physical education, fine arts, and personal planning. Although the curriculum in BC was revised in 2016, these subject-specific learning outcomes remain and are specific to grade levels and separate subjects. It does, therefore, emphasize discipline-centred learning and encourage age-specific achievement levels in each of the content areas. Rather than promoting an ecological view, this curriculum shows characteristics of what Sterling (2001) has described as a mechanistic educational paradigm in that it is prescribed, detailed, and largely closed; incorporates de-contextualized and abstract knowledge; and focuses on subject disciplines.

In deciding to model and teach sustainability, this school tried to infuse instrumental sustainability outcomes of reducing their electricity by 40%, waste by 40%, and water by 10% across the curriculum with the help of the Sustainability Co-ordinator, who organized special workshops and offered to help teachers integrate it into their curriculum. As external funding paid for the Co-ordinator's position, this dictated the time needed to be spent on measurable outcomes in reducing electricity, waste, and water, thereby limiting the scope for further curriculum development, beyond an instrumental orientation. In retrospect, having to rely on funding sources that require measurable outcomes or not having expertise in incorporating curricular development or process needs into funding proposals was identified as a significant obstacle. When asked if curriculum development would have been funded, as teaching sustainability had been a goal, the Principal replied:

> It may not have been. [The funding NGO] would not have funded someone writing curriculum but we still might of gotten around that writing into the proposal a little more detailed how we were going to reduce water, energy and waste: involving kids, writing curriculum. They might have had to be above and beyond the grant like all the planning, preparation teachers do after teaching hours.

This last point of involving teachers in incorporating sustainability into the curriculum, however, was restricted as teachers were told it would not entail an increased workload. A teacher and the Community Board Chair felt the grant process had structured what they were doing:

> It has been a very limiting factor. Lots of ideas would come through the year and we wouldn't be able to follow them up as we needed to fulfill the grant obligations first and we only have so much time.

This quantifiable, instrumental focus is resonant with the mechanistic educational paradigm (Sterling, 2001). In addition to focusing on measurable outcomes of reducing waste, water, and electricity, while not incorporating curriculum development, the grant also limited the amount of time the Co-ordinator could put into the Program. As the Principal noted, "The Coordinator's position is very limited at only two days a week. We are very limited in the amount of time that she had, given the breadth of the grant." As a result, much of the sustainability curriculum was instrumentally focussed on the Co-ordinator giving lessons in the Grade 6/7 science classes about changing behaviours to reduce water, waste, and electricity and trying to organize some initiatives such as recycling and Earth Week presentations or activities to support that.

A parent and teacher at the high school level felt the provincial curriculum was an obstacle, set up as it was with learning outcomes in separate subject disciplines. As she explained,

> I don't know if you can be much of an innovator when the rest of the system is looking this way. It is like a felt board, stuck on top of the school system, and what I would like to see it be is something completely stand-alone and completely stir the pot up. To really make its mark it needs to be interdisciplinary, it just has to be.

As the teachers were focused on the government subject-centred curriculum, the sustainability programme was extra-curricular. Some teachers could see where sustainability could be integrated if they had time and that was mandated, but they also identified one of the main obstacles was not having a clearly defined sustainability curriculum and scope and sequence that would fit into the mechanistic teaching structure. Being immersed in, and used to the mechanistic system, they expected sustainability curriculum to be planned and imposed from the top-down, with resources, so they could simply teach it. Although some teachers saw the potential to integrate sustainability into various subject curricula, one of the teachers said,

> I stick pretty much to my curriculum just because I feel haggard. If I can open a book that says teach this in math today I'm relieved. It's a lucky break for me. I'm not about to reinvent the wheel if I am under the gun time-wise; Government of Education wise. It might not be my philosophy but there are certain times I just go, "Tell me what to do."

Outside of this case study school, where teachers were reluctant to move beyond their subject boundaries to integrate subject-based curricula in their own practice, a practicing administrator, Nicol Suhr, a graduate of the Master's in Environmental Education and Communication program at Royal Roads University, found teachers actively trying to stop another teacher from implementing an integrated outdoor curricular programme:

> Upon intentional integration of outdoor learning activities in an introductory grade 8 program (of approximately 3 weeks per group of 30 students), one teacher successfully grew an outdoor education program into grade 8, 9, 10, and 11/12 classes. The next step was to develop a specialized grade 10 cohort with a semester long integrated science/English/outdoor education/physical and health education class. With the support of the Administration, this was communicated to the staff, students, parents, and included in the course selection booklet, with successful cohort numbers requesting the program. However, once the numbers of students interested were released, teachers not involved with the program protested the teacher leading the program as he had not previously taught science 10, despite being qualified. Union representatives were involved, and teachers attempted to utilize collective agreement language around staffing and seniority to limit the outdoor education teacher's permission to teach the science aspect of the course. Upon departmental arguments and union involvement, the outdoor education teacher withdrew from the program and the program was dissolved even before the timetable was completed.

In this case, although there were efforts to integrate the government curriculum with outdoor environmental education, the *defence of boundaries* in maintaining the traditional mechanistic curriculum structure and *competitive, survival of the fittest* undermined this innovative effort.

Teaching

Teaching is also imbedded with mechanistic root metaphors of the industrial paradigm. *Society is best controlled when power is centralized, progress is linear, take the world apart to understand it, and competition, survival of the fittest* are all very obvious.

Centralized Control

As we saw in the government-run, case study school, some teachers are trying to be innovative, but teaching in a mechanistic paradigm is also based on top-down control, focused on the prescribed curriculum. Based on the mandated curriculum, the teacher controls what, when, and how to learn, and determines if the students have learned.

In this linear, top-down system, teachers are often reduced to technicians, with their priority to teach the prescribed curriculum from textbooks and manuals. When asked what you teach, teachers typically respond with a grade level, such as "I teach Grade 3" at the elementary level, or "I teach Science" for example for the high school grades. They rarely say they teach students.

When speaking to practising teachers involved in my curriculum design course for a Master of Education in Sustainability, Innovation, and Creativity, they could see a number of ways centralized control was exemplified in their own teaching and the impact it had on students.

Teacher A: The ministry of education has control over what curriculum is taught in each grade but then that is also further controlled by the teacher who decides in what way/how the curriculum will be taught.

Teacher B: In a classroom setting, a teacher is traditionally or most commonly seen as the leader of the classroom, who knows everything, and that knowledge is transferred from the teacher to students. The teacher performs classroom management to control the class to meet classroom expectations, a teacher gets to decide what grade the students deserve, how students learn, and so on.

Teacher C: I would say that in many classes the power within the classroom is centralized with the teacher. Many teachers still use direct instruction where the power to control the discussion/instruction is kept by the teacher. As well, it is often the teacher that has control over assessment. Instead of allowing students to self-assess and give input about what they feel they have learned.

Teacher D: This centralized power was how students were trained to become obedient to power. This centralized power, however, is not how democracy functions in modern society.

Students have very little choice, if any, and are not seen as partners but recipients in a system dominated by higher authorities and experts who are presumed to know best.

But teachers also felt their teaching was affected by centralized control at higher levels. A number of teachers voiced concerns over assessment as well as professional development being centrally controlled, as represented by this quote, "I see it often when a school division decides that only one type of assessment method will be used for certain subjects and that professional development is determined from the top down." Another teacher unexpectantly experienced frustrations of centralized control, sharing,

> Just this morning, I am hearing that there will be another schedule change made to our school day. Two classes are to merge into one, as the numbers are too low! It was making me think again about this metaphor at play! The top-down hierarchy with students at the bottom having zero voice. Due to a lack of cafeteria supervision, the classes will be merged so that the teacher can supervise the cafeteria! Talk about a waste of resources. I am cranky about it but this further supports that the top-down approach still exists. The student and the teacher have little say in how they learn or who they learn from.

Progress Is Linear

Another influential root metaphor that is prevalent in our educational system is that progress is linear. In the traditional linear system, teaching is based on transmission of knowledge through cognitive experiences. Although teachers are expected to take students' individual needs into account, the system is not designed to cater to them beyond a very superficial level. Given the emphasis on analytical thinking and the linear thinking it relies on to assemble separate pieces of information, teaching has become centred on knowledge and learning is structured on progressive building blocks. As a result, teaching is dominated by a transmission of preprogrammed blocks of instruction through a curriculum scope and sequence that states what to teach and when. A practising teacher identified how this was active in their work,

> As teachers, we are expected to teach a designated curriculum for our specific course or our grade level. When the curriculum has been

covered and students have met the outcomes they are passed along to the next grade/course. By design, our education system is set up similarly to an assembly line: we group students of similar ages together and expect they will all be able to meet the learning targets within the given time frame. Students need to learn certain concepts before moving on to the next level of learning, however, the assembly line approach does not account for diversity in learning and makes assumptions that all students within the same group are ready to learn at the same time and pace. The linear assembly line approach equates learning to a straight-line path from acquiring knowledge to producing an outcome-based product.

Another practising teacher recognized linear thinking also influences a teacher's approach to assessing the learning,

> Assessment follows the same idea. Often a unit is assessed and then you move on to the next unit, rarely coming back to it. This is detrimental if something clicks for a student later in the year for a previous unit or topic.

Such an approach is vulnerable to teaching that focuses on transmission of knowledge, passive learning, and noncritical inquiry. Teachers are caught in a dilemma. Although progressive teacher education emphasizes the need for child-centred learning, innovation, and creativity, within each separate subject discipline, teachers have been trained in a mechanistic paradigm and are still expected to teach to predetermined learning outcomes that are to be sequentially learned from kindergarten to graduation through an efficient transmission of curriculum content (BC Ministry of Education, 2015; Ontario Ministry of Education, 2019; Robinson, 2015). This mechanistic, linear system is a barrier to creative, transformative, child-centred learning.

Reductionism

Heavily influenced by Descartes and the age of Enlightenment Reasoning, schools teach us to separate the analytical mind from our emotions, from our creativity, from our bodies, and from Nature (Orr, 1992). As we saw earlier, the left-brain emphasis in schools is to take things apart to understand how they work, to focus on the details, so we can then predict and control our world and ourselves. The majority of time teaching in schools is spent on developing and emphasizing the skills of the analytical left brain. Teaching methods emphasize the left brain's analytical, critical thinking skills, separating learning from its context, yet it is learning in context that connects left and right brain thinking, the rational and emotional, giving meaning to the subject (Lakoff, 2010). We also teach kids to separate mind and body: learning about the body in science is rarely if ever connected to physical education classes because one takes place in a classroom and the other in the gym, typically taught by different teachers and guided by very different learning outcomes. In schools, the precise, analytical subjects of science and mathematics are prioritized and receive higher status than the creative, intuitive subjects such as art or music associated with the right brain (Robinson, 2015). And when budgets are cut, it is the creative, intuitive, context-based subjects, those that emphasize the right brain, and the connections between the two hemispheres and the body that are dropped (Goodlad, 1984). Two practising teachers in my curriculum design course spoke to this.

Teacher A:	Many educators believe that the sole purpose of the school system is the academics and the ultimate truths. I have seen educators say things like they do not need to teach social emotional learning or sustainable education as those are "soft skills" and not the purpose of education – that is to think about our learning and to discover facts in a black and white way.
Teacher B reported:	The subjects are all compartmentalized and there is a massive emphasis placed on the core subjects of math and literacy (the left-brain, logical topics). The right-brain topics are often neglected and sometimes even cut from the educational system; students' intelligence is measured by their abilities in those core subjects. In our school board, we don't even comment on art or health on the report cards, as if they don't matter.

Survival of the Fittest Based on Individual Achievement

As learning is product oriented in today's Western educational systems, evaluation of student achievement is central to sorting and categorizing learners as they move through the system (Senge et al., 2012), with teachers responsible for assessments and report cards to report on subject discipline progress. As noted earlier, to know if teaching has been successful, we have relied on testing: international comparisons, external inspections, national standardized testing, provincial exams, and classroom assessments. Conformity rather than diversity is valued as all students, regardless of their differences, are all given the same test. This supports the competitive mechanistic root metaphor of *survival of the fittest* based on individual achievement. We reward those top students with gold stars, stickers, top marks, and/or public recognition through award ceremonies and certificates of merit. Although effort and anecdotal comments may be included in report cards, students and parents want to know how they have done in relation to others. Competition for those top marks are powerful drivers that enable further education and scholarships. The system is not designed to celebrate our diversity, our ability to collaborate, our ability to recognize the influence of various aspects of complex systems as they interact, our creativity, or our ability to innovate. As Ken Robinson (2015) notes, our industrial approach to education has focused on efficiency and yield – on maximizing the transference of knowledge rather than on creating a rich environment where creativity and learning can thrive.

This clearly frustrates many teachers who try to work counter to this system, trying to teach students, encouraging creativity, systems thinking, and innovation. Innumerable valiant efforts through the 1960s to the present, to develop creativity, environmental, experiential education, and ESD, continue to be subverted by this system that is not designed to support nonlinear transformative approaches and perspectives (Ireland, 2007). Typically, more transformative, student-centred approaches such as experiential learning or the creative Genius Hour are given a slot at the end of the day within the mechanistic teaching schedule. Throughout my curriculum design courses, teachers report trying to use student-centred, transformative pedagogy, but report being limited by both the timetable breaking the day up with other subject specialists and being responsible for reporting on predetermined learning outcomes and competencies in separate subject disciplines.

We now know from neuroscience that our right- and left-brain hemispheres actually work together with our emotions and whole body in relation to our environment (Lakoff, 2010). But our curriculum and teaching, and particularly evaluation, are not set up to support this type of learning.

Systemic Effects

Although the teachers agreed to model and teach sustainability in the case study school, there were significant issues rooted in the mechanistic system the teachers are caught in, as well as their own embodiment of mechanistic root metaphors of *society is best controlled when power is centralized, I think, therefore I am (cognitive, analytical, left-brain emphasis), take the world apart to understand it,* and *knowledge is built of progressive building blocks.*

The Principal recognized they still had improvements to make in their teaching methods. Interviews with the teachers and parents revealed an instrumental, transmission view to be the most prevalent. The educational focus was product rather than process oriented, emphasized centralized control in teaching, and focused on functional rather than critical and creative competence.

Many Intermediate teachers and parents commented that much of the sustainability teaching was at the level of transmission of knowledge, which Miller (1996) associates with a mechanistic view rather than transformative pedagogy, in line with an ecological view. The Program was imposed rather dogmatically as it was designed from the outset to meet reduction targets and students were not involved in this decision-making process. The Grade 5 Teacher recognized that a transmission approach was commonly used at the school:

> I think in terms of sustainability what we do in the schools is limited to teaching kids about reducing, reusing, recycling. Environmental education in the schools seems to be divided between putting our things in the recycling depot versus throwing it in the garbage because it is easier, and on the other side love the outdoors, smell this tree, don't you feel wonderful. The whole space in between which is everything else in terms of taking ownership of the creek bed, and saying that's our creek and we rebuilt it; and those are our fish because they are coming back. That sort of ownership in nature I don't see happening in schools.

Although this teacher recognized sustainability education could be so much more, it was interesting to see the anthropocentric view, suggesting humans should have a sense of ownership over Nature.

The Grade 6/7 students felt the transmission approach created negative feedback that affected their interest and participation. This was particularly evident when these students were asked about the recycling programme:

> If some teachers see you put recyclables in the garbage, they send you to the homework room or keep you in at recess. I don't think they should do that because then we just don't want to do it because we don't want to do what they've told us. They remind us 24/7 and it gets so annoying that we just disobey them.

In support of this, a parent felt her son's rejection of the Program was largely because it was imposed.

The Grade 5 Teacher was also frustrated with the programme being imposed, saying there had been little connection between the sustainability programme and what had been going on in her teaching and learning, honouring previous knowledge:

> I don't see much connection to teaching and learning. I think the first year they tried. I guess they were assuming everyone and the kids are at zero. That's a big assumption. Most of the kids had some great ideas. I have been recycling at this school for five years. We sell our Tetra Packs and that's how we help fund our fieldtrips.

In addition to this top-down plan and impose approach, a mechanistic educational style was most obvious as learning was structured through separate subject disciplines in all but the Kindergarten/Grade 1 classes. This was exemplified when the Program Co-ordinator taught sustainability classes to the Grade 6/7 students in science classes. In speaking about how she developed learning in sustainability she recognized the integrated essence of sustainability in saying, "When I taught the science curriculum to Grades 6/7 on sustainability, I would bring in social aspects of scientific considerations." But when she was asked if she was integrating the social studies learning outcomes (IRPs) with her science teaching she showed how heavily discipline centred her teaching was:

> I don't know because when I did the science with them I didn't teach the social studies so I'm not familiar with the social studies IRPs. But a social studies teacher saw social studies as perfect for teaching sustainability.

The Program Co-ordinator noted that cross-curricular concerns had not been discussed, as other teachers were not comfortable with sustainability and cross-curricular teaching. The Intermediate Team of Grades 5, 6, and 7 worked as a team to specialize in separate subject disciplines rather than teaching all the subjects in their respective grades. When the Grade 5/6 Teacher was asked if she brought sustainability into her teaching, she said that although she did bring it into her social studies classes to some degree, she showed how discipline centred her teaching was as well:

> It's happening, not really as part of my subject curriculum in terms of the recycling. I don't teach science so don't get involved with some of the science aspects. In my socials program we are talking a lot about social responsibility – it is everybody's air so we need to think about our actions and not using more than our share. It's been great letting kids know things can be done without having to use pesticides and poison the air and the Earth. Also, when you go Christmas shopping students could buy from a Ten Thousands Village Store so the profits go back to a developing country.

A Grade 6/7 Teacher who is part of the Intermediate Team felt there was very little integration with sustainability concepts:

> It depends on the teacher. It is difficult to tie it in. It is tied into the management side of the school in recycling. It is not incorporated into the learning. When I teach math I'm not teaching environmentalism, when I teach reading, maybe a little bit because you are thinking and writing. It varies with the group of students you are working with.

Even the Grade 4 Teacher who had her class make a video on water that focused on ecological literacy was very discipline centred. When she was asked if she integrated subject disciplines around themes, she replied:

> No. I'm not one of those thematic teachers in the primary sense where the whole room becomes a rainforest. I tended to be an Upper Intermediate before I was Grade 4 so I teach my subject areas separately.

When the Intermediate Team was asked why they don't integrate sustainability more into their teaching, they exemplified their mechanistic defence of boundaries and linear thinking saying they first needed to be provided with a scope and sequence on how to teach it. Also, they were not given any professional development, or chance to develop their insights through team teaching with the Program Co-ordinator. The classroom teachers were given teaching preparation time during the same time the Program Co-ordinator was scheduled to teach sustainability science to their classes, so, as a result, they were not present for any of the sustainability-focused teaching. This was underlined with a strong defence of boundaries by maintaining it wasn't their job, it was the Sustainability Co-ordinator's job.

Reinforcing the notion that humans are separate from the environment and learning is decontextualized, the Program Co-ordinator recognized most of the teaching and learning at the school took place indoors. The teachers felt more comfortable in a classroom and there were many difficulties controlling classes outside. When asked what percent of school lessons are outside, she replied it was not many as it was harder to teach outside, trying to contain 30 students without the four walls of a classroom.

Overall, the teaching approaches of cognitive, passive experiences, non-critical inquiry and a restricted range of methods have associations with Sterling's (2001) mechanistic model. The students in Grades 5–7 confirmed that most of their classroom experiences, either focused on the sustainability programme or otherwise, were based on cognitive experiences through passive instruction. Further evidence was gained in observing three of the Intermediate Grade 5/6/7 teachers as well as the Grades 4 and 3 teachers in their classrooms. The desks were all laid out in rows facing the teacher's desk or blackboard at the head of the class. The emphasis was on didactic teaching by way of passive instruction encouraging individual cognitive learning. The walls of the class displayed examples of this logical, linguistic work. Environmental content, if it was incorporated, was taught dogmatically through simple statements focusing on awareness rather than investigating concepts at a deeper level through critical thinking, systems thinking, and empowerment.

As a result, the sustainability programme was largely taught by the Program Co-ordinator in science classes for the Grades 6 and 7 students; or as a special one-off project in various grades for Earth Week, such as creating a Nature walk, a composting centre, mural, or a video on salmon and water in the few weeks leading up to Earth Week presentations to the community. There was very little integration with the government-prescribed curriculum.

Learning

For the student in the reductionist, mechanized educational system, learning is individualized, focused largely on logical-mathematical, and verbal-linguistic intelligences (Gardner, 1993; Goodlad, 1984). The main purpose is to absorb the prescribed curriculum and show functional competence. After successfully teaching oneself to walk, speak at least one language, and understand the social morés of their society in context, children enter school and are no longer honoured as critical thinkers and competent, self-empowered learners. From the moment they enter school in the linear, mechanistic system, until they graduate, they are told what to learn, when to learn, how to learn, and then if they have learned. When the bell goes it is time to stop thinking about science, for example, and start an unrelated social studies or language assignment – or be evaluated – whether students are ready or not. What is left for the student to decide? Students who try to make their own decisions as to what, when, or how they want to learn do not fit the system and are either forced out or are brought back in line, forced to conform (Goodlad, 1984; Hern, 2003). The student has little to no say about what, when, or where they learn or even if they have learned. This reinforces the root metaphors: *society is best controlled when power is centralized, the world works like a machine, take the world apart to understand it, and survival of the fittest.* The practising teachers comments further exemplified mechanistic root metaphors at play in their students' learning.

The World Works Like a Machine

In applying the traditional mechanistic model, the individual gets lost as it is typically assumed that all children will learn the same thing, in the same way, at the same time. It is a very instrumental, mechanistic process designed to fulfil predetermined outcomes through efficiency, control, productivity, and yield (Robinson, 2015). In my work with teachers who are currently practising in the mechanistic system, the following quotes from four different teachers, from elementary school to high school, represent how they see the educational system functioning like a machine.

Teacher A: A machine follows certain ways of doing things because of how it was coded/designed, and we know what we put in is what we will get out. This is how education will try to be ....

Put all of the kids who were born in the same year and in the same community into the machine, give them the same directions, and expect the same product at the end – workers who are turned out at age 18, although this clearly does not work.

Teacher B: I find that this machine-like model often does not provide the flexibility diverse learners require. Often in academic institutions it feels as though you are checking a series of boxes to get to the next level instead of learning based on passion and engagement.

Teacher C: Students need to jump through certain hoops to ultimately get to the next level – much like a video game. Sadly, this machine doesn't stop for anyone and often students who struggle with certain skills or encounter any number of obstacles get pushed through despite being able to take on more difficult concepts. I think of the semester system in high school where students have a limited amount of time to show their learning in order to pass the course or they have to retake the course. I remember feeling like Math in high school felt like a wild ride where I could never quite keep up or understand a concept before a new one was introduced.

Teacher D: Depending on the subject, it almost encourages students to completely forget what they've learned and move on to the next topic. Thinking back to my social studies classes as a student, I never once thought how different time periods affected each other. As one unit ends, I'd just push all that learning aside and focus on what's new, never thinking of putting it all together.

Competition: Survival of the Fittest

Individual achievement is typically seen in relation to others so that competition for achievement and recognition can overshadow personal growth, satisfaction, and positive self-esteem. Our educational system fosters a real fear of failure, which in turn inhibits growth and expression (Goodlad, 1984). Practising teachers in my 2022 Master of Education course identified a number of ways competition, or *survival of the fittest*, is prominent in their schools:

Teacher A: High-achievers are often rewarded and praised, whereas less successful students sometimes give up as they cannot compete.

Teacher B: Oftentimes, students who perform the best in sports are the ones who are recognized with awards and those who do not yet possess the skills are cut from teams. Children who can't keep up with the intense demand in some courses either drop the course, fail or fall through the cracks in various ways.

Teacher C: We see similarities within our IB [International Baccalaure-
ate] Diploma program too. Those students who are deemed
bright, are taken to the lab more frequently and have more
enthusiastic teachers. Through increased lab time, they gain
more practical skills that will help them as they progress
from Grade 11 to Grade 12 and onto University. They
are given better instruction by teachers who see value in
working with them, and as a result, get more time poured
onto them for extra help and conversation to check for
understanding.

Survival of the fittest is also dominant in assessment. As a practising
teacher noted:

Teacher A: I also see this also our current grading system. Much of our
grading system breeds competition, especially when schools
say that there can only be a certain number of As or 4s (whatever system it is using). Those who get to the top get special privileges associated with honour role and are continually
praised for their grades. Not that we should not celebrate kids
who work hard, but often we only look for those top finishers
and celebrate them.

Teacher B: Teacher B recognized this competitive orientation was also ex-
hibited by parents when their school shifted from letter grades
to descriptors of progress, "[Parents] also wanted to be able to
rank their kids and I think that they struggled to understand
what *extending, proficient, developing* and *emerging* meant.
Many of them asked if an extending is like a A+/A."

Other teachers felt competitive assessments really influenced learning.
One teacher was thinking about her students by saying, "I feel they have
been programmed to learn a specific way and have gained the necessary
skills to perform well on tests." Another teacher looked beyond the classroom assessments to standardized tests they were required to administer
and reflected on how it affected teaching and learning:

Training for these high-stake standardized tests also stops all thinking,
discussing, and community building as the teacher has to spend time
preparing students for the test. The things a high-stake standardized
test can measure as an assessment tool is narrow. It also sacrifices opportunities for students from having chances to explore their creative
potential to be able to find creative solutions for future obstacles they
may encounter for the world they live in.

These findings are significant as they highlight how prevalent the mechanistic root metaphors are in education as well as their detrimental implications. As illuminated by practising teachers, education has continued to be dominated by a left-brained cognitive emphasis; enlightenment reasoning; linear thinking; reducing wholes down to their separate parts; treating individuals as cogs in a machine, trying to force square pegs into round holes even when trying to model and teach sustainability.

Systemic Effects

The impacts of the mechanistic educational paradigm and centralized control were keenly felt by the students trying to learn about sustainability in my case study research. An Intermediate student referred to the fact that learning at their government-run elementary school was very teacher focused with little student empowerment. Although there was an independent learner programme, this student was not able to get permission to learn in this way:

> I am stuck in the class that has to go along with the teacher because you have to get 80% or better on tests to be in the Independent Learner class. But I think I'm much better at learning individually. I completely lock up for tests.

We saw in the "Teaching" section, learning was focused in separate subject disciplines, often with different teachers, based on a linear progression through the government-imposed learning outcomes for each separate subject; with sustainability being learned through a science education lens.

Any other learning in relation to sustainability took place through extracurricular activities, but these opportunities were also for the Student Advisory Committee for students in Grades 6 and 7. In frustration, a Grade 5 student finally started a Kids' Club, with the help of their mother and the community school so kids of all ages could be empowered to learn about and develop their own sustainability initiatives, based on their interests. Interestingly, after two years, the school's focus on sustainability ended when they lost funding for the Sustainability Co-ordinator and the Principal was transferred. The only thing that remained was the Kids' Club, run by kids.

So Why Does This Matter?

The industrial foundation of education, illuminated by root metaphors of the world (*the world works like a machine, nature is a resource, centralized control, take the world apart to understand it, progress is linear,* and *survival of the fittest* (Schwartz and Schwartz, 1995; Sterling, 2001), clearly frustrates efforts to develop systems thinking, innovation, and creativity, as the system is not designed to support nonlinear transformative teaching approaches (Robinson, 2015; Sterling, 2001). The formal educational system does not have a history of supporting empowered holistic learning necessary for systems thinking and life-long learning in developing sustainably. The mechanistic foundation completely disregards the environmental context that life and learning take place in as well as the holistic Nature of children who learn through all their senses and through social and socio-ecological networks. Education is seen as happening in school buildings, fenced off from their communities and surrounding environments.

After at least 12 years of being trained to think in bits and parts, being trained to think in separate subject disciplines, we rarely see how science, social studies, language arts, and mathematics are connected. Or how working with the head, both creative and analytical, the heart and the hands in concert with those all around us in a meaningful context can lead to greater understanding. Students tend to know a lot, if they can recall the information, about various bits and pieces of decontextualized information and very little about how it all fits together and has meaning, like trying to do a jigsaw puzzle, and fit all the pieces together without a picture to guide how they relate to each other.

This way of thinking and educating may have been acceptable when Industrial society needed many well-trained graduates to become well-trained workers to keep the assembly lines of the economy going based on predictable job markets. The problem is that students and the world aren't stable and predictable for our industrial educational approaches to be effective – ask any teacher about classroom management and student control issues they face on a daily basis. We did not predict the impact of the digital revolution, the internet or the development of artificial intelligence, new pandemics, climate change, or biodiversity loss. Nor can we predict and control our environment as increasingly devastating impacts and occurrences of hurricanes, drought, floods, and wildfires remind us.

In reality, the world – including each student – is unpredictable, constantly changing, and adapting. We do not know what jobs will be needed going forward in our rapidly evolving technological society, yet we continue to prepare students for jobs that are rapidly changing or disappearing. From a mindset of thinking we can control things, we tend to see this as a threat. Rather than ignoring the messages and feedback from students, our evolving society, and our surrounding environment, we need to be attuned to it, see the patterns that connect, and learn from what is emerging. We need divergent, innovative, systems thinking. Once we recognize and educate for this, we will be far better prepared to anticipate and handle change in a positive, innovative way.

Creativity and divergent thinking are not developed through a transmission of predetermined learning outcomes. When we teach about a problem by focusing on one aspect in isolation, mechanistic thinking makes it really hard to understand, much less anticipate, how this has a greater impact on another part or on the system as a whole. When chlorofluorocarbons (CFCs) were used in aerosol sprays and refrigerants, no one had any idea, or could even imagine, they could create a hole in the ozone layer of our atmosphere. The same goes for unanticipated challenges we are now facing in social injustice, deforestation, loss of biodiversity, and climate change ... these are symptoms of an unsustainable, mechanistic worldview and the actions that support it. David Orr, in 1994, noted that the destructive effect of globalization and development upon the climate, ecosystems, and traditional cultures "is not the work of ignorant people. It is, rather, largely the result of work by people with Bas, BSs, LLBs, MBAs, and PhDs" (Orr, 1994, p. 7).

Conclusion

As we've seen, creativity, diversity, and unpredictable learning outcomes are not supported by the industrial philosophy that initially saw the student as a blank slate for a transmission of knowledge (Robinson, 2016). Yet creativity, diversity, and being able to innovate are exactly what we need if we are to redesign our society to become sustainable. Divergent thinking – the ability to think on one's own, to generate ideas by exploring many possible solutions, beyond the norm – is the basis for, and a key component of, creativity. A longitudinal study, published in 1992, shows how divergent thinking tends to decline with increased schooling. This study found that out of 1600 children, 98% entering school at age 5–6 were divergent thinkers. Those same children by age 8–10 showed only 30% were still divergent thinkers and by age 13–15 only 10% were. The study then tested 250,000 25-year-olds and found only 2% were divergent thinkers (Ainsworth-Land and Jarman, 1992). This research shows that the school system and how we educate overwhelmingly encourages learning to conform, effectively discouraging our natural tendency to learn by thinking and considering creative, divergent options. In reality, our social framework and our educational experiences condition the way we think – often without our realizing it.

John Taylor Gato, New York State Teacher of the Year in 1991, emphasizes the system has not changed from its historical roots of trying to control the masses; and he's now one of the most vocal opponents of our public educational system. He emphasizes there is a hidden curriculum of command and control that denies individual creativity, expression, and decision-making, imposed, even by the most highly decorated teachers, due to the way schooling is structured. In this excerpt from The Six-Lesson

School Teacher, Lesson 4, Gato speaks of this hidden curriculum and those students who fight the system:

> Curiosity has no important place in my work, only conformity. Bad kids fight against this, of course, trying openly or covertly to make decisions for themselves about what they will learn. How can we allow that and survive as schoolteachers? Fortunately there are procedures to break the will of those who resist.
>
> (Gato, 1991, para 11/12)

But what about those students who do not "fit" the system, or those who do fit? Those who "fit" become channelled into perfecting, and further promoting this mechanistic way of thinking. Those who don't fit as they move along the system "fail", may be labelled as "learning disabled," "special needs," "troublemakers," or having attention-deficit hyperactivity disorder, often controlled with drugs so they can cope in a rather restrictive, narrow parameters (Louv, 2005; Robinson, 2015).

My experience as a K–12 teacher for 30 years is that many students cope with the system that has failed them personally by performing indifferently, not cooperating, not doing the work assigned, or in many cases, not attending. It's the teacher's job to keep these students in line and to try to teach them, and the administrator's to further discipline and find ways to deal with them when all else fails. In short, the system is structured for efficiency to control the greatest number of students from age 5 to 18 while transmitting the greatest amount of knowledge. It is not structured to support student diversity, creativity, and contextual, transformative learning.

Although educators are trying their best, defending the system as trying to educate and empower the masses, as we've seen, it is still very machine-like and centrally controlled through a rigid top-down hierarchy, with kids on the bottom and parents on the outside fringe having little or no say in what or how students learn (Goodlad, 1984; Ireland, 2007; Robinson, 2015). Trying to work effectively in this system can be very frustrating, lead to teacher and student burnout, and result in various power struggles at all levels between students and teachers, students and parents, parents and teachers, and teachers and administrators. Everyone is vying for control and defending their territory, while trying to get their needs met (Ireland, 2007). It is not a picture of teamwork, coevolving for educational effectiveness in a changing society.

We live in a complex society where we need creativity and innovation and the ability to think systemically to understand and deal with complex issues arising in complex socio-economic-ecological interactions. Many such issues are threatening social justice and our very life-support systems that provide us with clean air, water, and healthy food. Our traditional system of education is not effective for educating children to realize their whole potentials: to be part of and positively contribute to a changing environment, society, and economy with creativity and innovation; understand how things interact in complex systems; contribute through jobs that have not yet been invented in our rapidly changing technological society; and come up with new solutions to problems society doesn't yet know how to manage. We need education to be effective to engage in life-long learning, so students of all ages are motivated and able to respond to the changing needs of society.

Yet school tends to focus on educating students about complex reality, such as climate change, for example, by continuing to break it apart into separate units of study and pieces of information sprinkled here and there in science or social studies. This does not deal with the complex nature of these issues. It does not prepare students to understand and deal effectively with our complex changing world and our role within it. And continuing along the assembly line from public school, simply adding more vocational careers, and specialized courses in universities does not address our need to understand systems and how to live better in a sustainable manner as a peaceful society. Instead, it perpetuates the unsustainable mindset of the status quo.

With a mechanistic system based on centralized control and top-down hierarchies, the lower you are on the hierarchy, the less empowered you are to see or make connections or contribute to our evolving systems. The Ministry of Education controls budgets and mandates curriculum, prescribed learning outcomes, standardized testing, and graduation requirements. The College of Teachers controls teacher certification. The school board is responsible for Ministry policy implementation. Superintendents control staffing, district budgets, and infrastructure like buildings and transportation. Principals control school budgets, staffing, class configurations, and student compliance. Teachers translate prescribed learning outcomes into lessons, direct and evaluate student learning, and prepare students for standardized tests. Students are required to be compliant and learn as directed.

Education encourages us to take this mechanistic way of thinking with us and use it with everyday things we are involved with – hence, it perpetuates our problems. In society we've treated the natural world like an unlimited resource to be exploited for unlimited growth; we throw things "away" as if we are separate from Nature and there is such a place; we often see things simplistically as cause and effect; we analyse our problems but rarely synthesize; we typically think that an awareness of all the details will automatically lead to action to solve all our problems, while not understanding complex interactions; we plan and impose from the top-down, managing through command and control, yet expect our citizens to be empowered and engaged (Senge et al., 2012). This is the mindset of the past industrial unsustainable society and it is imbedded in, and perpetuated by how we have been educating ourselves with schools modelled on factories that educate children to be cogs in the machine of society. From how we teach, where we teach, and what we teach – to how we administer education – we are teaching students that *the world works like a machine* where things can be best understood by breaking them down into their separate parts. This industrial approach to schooling is inhibiting our ability to create a sustainable future as it is not equipping us for the challenges of the 21st century. As Ken Robinson (2015) notes, our industrial approach to education has focused on efficiency and yield, on maximizing the transference of knowledge rather than on creating a rich environment where creativity and learning can thrive.

Even though our present society has been heavily influenced by the mechanistic views of Descartes and Newton, the world doesn't really work like a machine and people are not individual, independent cogs. In the early 1800s von Humboldt recognized the interdependence of all things (Wulf, 2016), but although he was highly influential in his day, the reductionism of Descartes and Newton dominated. Yet since, there have been further significant scientific developments in the 20th and 21st centuries that have extended von Humboldt's theories and our understanding. In physics, at the subatomic level, the tendency for matter to exist is influenced by the complicated web of relations between the various parts of the whole – even the observer is seen to influence what is observed (Capra, 1996). Beyond physics, new discoveries in ecology, biology, fractal mathematics, and neuroscience have also helped us understand our natural world and our relationship to it and each other as more holistic and interconnected (Noble, 2006). James Lovelock (1991) described planet Earth as a holistic, self-regulating system.

As such, the natural world is far more than a collection of separate species and natural resources. Trees are not simply roots, a trunk, branches, and leaves but are interdependent habitats, providing homes and nutrients, continually interacting with the air, soil, bacteria, the water cycle, and the sun, capturing CO_2, providing the world with oxygen, and communicating through the underground mycelium network in creating a biodiverse, interdependent community (Simard, 2021). Similarly, people are more than the heart, brain, and limbs. We realize students are complex beings, greater than the sum of their parts – but we don't educate this way. As complex beings, we are constantly affecting each other, interacting with natural systems, other cultures and environments with creativity and imagination. Concentrating only on the parts cannot understand or explain these interrelated complexities and does not help students reach their potential – able to adapt to our complex, dynamically changing world.

Although our scientific understanding has moved on, our educational system hasn't. Schools and the way we educate do not seem to have been significantly altered by scientific developments in understanding complexity of how the world works and how we are interrelated. We have heard talk of "participative learning" and "whole person education" since the 1960s, and even earlier. But these concepts have been simply graphed onto an educational structure that is designed to break issues into parts and address a "whole person" by educating various parts through separate art, science, maths, language, music, vocational, and physical education classes: their left brains in science, body in physical education, hands in shop classes, right brains in art classes.

The result is to subvert innovations and maintain the status quo. "Participative learning" challenges centralized control and prescribed learning outcomes and doesn't work well in "efficient" classes of 30+ students designed to transmit knowledge or achieve predetermined learning outcomes and high scores on standardized assessments. We recognize the need to reform education, however, without changing the system itself we are only paying lip service to such reforms, giving the illusion we are implementing significant changes. Very little change will come unless we recognize deeper mechanistic structures that need to be addressed. Standardized testing, rigid timetables, age-specific prescribed learning outcomes may seem like efficient ways to educate the masses or good ideas in isolation, but we need to see how these mechanistic structures influence the bigger picture; how they impact our thinking, learning, and worldviews.

The combination of linear, anthropocentric or human-centred thinking, and the emphasis on mechanism and reductionism influenced the underlying educational philosophy, the organizational structure and administration, buildings, grounds and resources, curriculum, teaching, and learning. We need to recognize education has been founded on and influenced by an outdated model and understanding of how to live effectively in the 21st century. Our educational system is a system in crisis. Rather than being part of the solution, we've seen that schools are actually part of the problem as they continue to promote the mechanistic mindset of an unsustainable worldview.

Knowing how our system of education does not question or challenge these assumptions and influential root metaphors of the industrial mechanistic worldview but incorporates them into administration, buildings, grounds, and resources, curriculum, teaching, and learning, helps us understand how education needs to change across all aspects of education. This is necessary to address the significant challenges we face if we are to transition to helping students develop the mindset and competencies needed to live and work sustainably in the complexity we find ourselves immersed in. The whole structure of education and schools needs to reinforce the ecological principles of sustainable living systems rather than the dominant, mechanistic system we inherited from the Industrial Revolution. If it does not, individual efforts on the part of teachers working in an atomistic system will tend to be subverted by the "hidden" curriculum, with its powerful, dominant metaphors (Orr, 1994).

This is not how *sustainable* living systems function. Fragmented, disciplinary thinking does not help students see the big picture, the essential connections, and interdependence of all these different disciplines and perspectives. We need to help students understand the world that works as a dynamic, interdependent whole; a system of *interdependence* where constant *feedback, diversity, adaptation, and emergence* based on creativity, and innovation are central principles. We need to realize *the whole is greater than the sum of its parts.*

We obviously need to rethink how we educate. Our mechanistic frameworks and ways of thinking are limiting our ability to rethink and redesign our unsustainable society. As the UNECE (2012) recognized, we need a complete transformation of education through policy, administration, teaching, and learning based on holism and systems thinking. We have an opportunity to free education from its restrictive, limiting cocoon. Our scientific understanding has transformed; the needs of society are transforming; now education needs to transform to meet these challenges. To allow ourselves to think differently – to develop creative solutions in a complex interconnected world – we need creative, systems thinking. What if education engaged the whole child, fostering creativity, systems thinking, and innovation to effectively address the world's most complex problems? It's time to consider a renewed educational framework for the 21st century, one that supports transformation and a diverse, sustainable society with positive socio-ecological relationships where all can thrive. To enable this transition, we can look to applying the ecological principles of sustainable living systems, as these provide a solid foundation to support rather than subvert the innovations we need, across all aspects of our educational systems, so we can develop sustainably. It's time we consider the future of school.

References

Ainsworth-Land, G. T., & Jarman, B. (1992). *Breakpoint and beyond: Mastering the future—Today.* HarperBusiness.

Alberta Ministry of Education. (2011). *Framework for student learning: Competencies for engaged thinkers, ethical citizens with an entrepreneurial spirit.* Retrieved from https://open.alberta.ca/publications/9780778596479

Amrine, M. (1946, June 23). The real problem is in the hearts of man. *New York Times Magazine.*

Association of Waldorf Schools of North America. (2015). *Waldorf education.* Retrieved from https://waldorfeducation.org/waldorf_education

BC Ministry of Education. (2015). *Curriculum redesign.* Retrieved from https:// curriculum.gov.bc.ca/rethinking-curriculum

Bowers, C. (1993). *Education, cultural myths, and the ecological crisis: Toward deep changes*. State University of New York Press.

Boyles-Deron, R. (2006). Dewey's epistemology: An argument for warranted assertions, knowing and meaningful classroom practice. *Educational Theory, 56*(1), 57–68.

Capra, F. (1982). *The turning point. Science, society, and the rising culture*. Bantam.

Capra, F. (1996). *The web of life: A new synthesis of mind and matter*. Flamingo.

Curtis, A. (2002). *The century of the self, a 4-part BBC documentary*. Retrieved from https://www.bbc.co.uk/iplayer/episodes/p00ghx6g/the-century-of-the-self

Dewey, J. (1938). *Experience and education*. Macmillan.

Gardner, H. (1993). Multiple intelligences go to school: Educational implications of the theory of multiple intelligences in readings and cases. In A. Woolfolk (Ed.), *Educational psychology*. Allyn and Bacon.

Gato, J. T. (1991). The Six-Lesson Schoolteacher. Retrieved from https://www.cantrip.org/gatto.html

Ghanaian National Council for Curriculum and Assessment (NaCCA). (2019). *Development of the Ghanaian curriculum*. Retrieved March 2023, from https://nacca.gov.gh/curriculum/

Gilead, T. (2005). Reconsidering the roots of current perceptions: Saint Pierre, Helvetius and Rousseau on education and the individual. *History of Education, 34*(4), 427–439.

Goodlad, J. (1984). *A place called school. Prospects for the future*. McGraw-Hill Book Company.

Government of India. (2020). *National education policy 2020*. Retrieved March 2023, from https://ncf.ncert.gov.in/assets/uploadedfiles/documents/NEP_Final_English_0.pdf

Greig, S., Pike, G., & Selby, D. (1989). *Greenprints for changing schools*. Kogan Page/World Wide Fund for Nature UK.

Hern, M. (2003). *Field day: Getting society out of school*. New Star Books.

Ireland, L. (2007). *Educating for the 21st century: Advancing an ecologically sustainable society* [Doctoral dissertation, University of Stirling]. STORRE. http://hdl.handle.net/1893/240

Lakoff, G. (2010). Why it matters how we frame the environment. *Environmental Communication, 4*(1), 70–81.

Louv, R. (2005). *Last child in the woods: Saving our children from nature-deficit disorder*. Algonquin Books of Chapel Hill.

Lovelock, J. (1991). *Healing Gaia*. Harmony Books.

Manitoba Ministry of Education. (2022). *Curriculum kindergarten to grade 12*. Retrieved from https://www.edu.gov.mb.ca/k12/cur/science/index.html

McInnis, N. (1982). A decade of environmental education. *School Science and Mathematics, 82*, 95–108.

Miller, J. (1996). *The holistic curriculum*. OISE Press Inc.

Noble, D. (2006). *The music of life*. Oxford University Press.

OECD. (n.d.). *National or regional curriculum frameworks and visualizations annex*. Retrieved March 2023, from https://www.oecd.org/education/2030-project/curriculum-analysis/National_or_regional_curriculum_frameworks_and_visualisations.pdf

Oelkers, J. (2002). Rousseau and the image of modern education. *Journal of Curriculum Studies, 34*(6), 679–698.

Ontario Ministry of Education. (2019). *Education that works for you.* Retrieved from https://news.ontario.ca/en/backgrounder/51527/education-that-works-for-you-modernizing-learning

Orr, D. (1992). *Ecological literacy: Education and the transition to a postmodern world.* State University of New York Press.

Orr, D. (1994). *Earth in mind: On education, environment, and the human prospect.* Island Press.

Republic of Kenya. (2019). *Basic education curriculum framework.* Retrieved March 2023, from https://kicd.ac.ke/wp-content/plugins/pdfjs-viewer-shortcode/pdfjs/web/viewer.php?file=https://kicd.ac.ke/wp-content/uploads/2019/08/BASIC-EDUCATION-CURRICULUM-FRAMEWORK-2019.pdf&attachment_id=&dButton=true&pButton=false&oButton=false&sButton=true#%5B%7B%22num%22%3A92%2C%22gen%22%3A0%7D%2C%7B%22name%22%3A%22XYZ%22%7D%2C69%2C228%2C0%5D

Robinson, K. (2010).Changing education paradigms. *TED Talk.* Retrieved from https://www.ted.com/talks/sir_ken_robinson_changing_education_paradigms

Robinson, K. (2015). *Creative schools: The grassroots revolution that's transforming education.* Penguin Books.

Robinson, K. (2016). *Sir Ken Robinson and Dr. Peter Senge: Education fit for the 21st century.* Disruptive Innovation Festival. Retrieved from https://www.youtube.com/watch?v=j1egR1szeH4&list=PLIGwcUQqMXyguvLdrkTazf4a8BniyVN0x&index=1&t=0s.

Schwartz, E., & Schwartz, S. (1995). Culture, ecology and education: The paradox of modernism in the transition to a postmodern world. *The Review of Education/Pedagogy/Cultural Studies, 17*(2), 167–174.

Senge, P., Cambron-McCabe, N., Lucas, T., Smith, B., Dutton, J., & Kleiner, A. (2012). *Schools that learn: A fifth discipline fieldbook for educators, parents and everyone who cares about education.* Cown Business.

Simard, S. (2021). *Finding the mother tree: Discovering the wisdom of the forest.* Penguin Random House.

Sjøberg, S. (2015). PISA and global educational governance – A critique of the project, its uses and implications. *Eurasia Journal of Mathematics, Science & Technology Education, 11*(1), 111–127.

Smith, G. (1992). *Education and the environment: Learning to live with limits.* State University of New York Press.

Sterling, S. (2001). *Sustainable education: Re-visioning learning and change.* Green Books.

Sterling, S. (2016). A commentary on education and sustainable development goals. *Journal of Education for Sustainable Development, 10*(2), 208–213. https://doi.org/10.1177/0973408216661886

United Nations Economic Commission for Europe (UNECE). (2012). *Learning for the future: Competencies in education for sustainable development.* Retrieved from https://unece.org/DAM/env/esd/ESD_Publications/Competences_Publication.pdf

Wulf, A. (2016). *The invention of nature: Alexander von Humboldt's new world.* Vintage Books.

Part III

Creative Solutions

You never change things by fighting against the existing reality. To change something, build a new model that makes the old model obsolete.

(Buckminster Fuller[1])

DOI: 10.4324/9781003389590-5

3 At the Crossroads
Guiding Principles of Sustainable Living Systems

As Buckminster Fuller said, "*You never change things by fighting against the existing reality. To change something, build a new model that makes the old model obsolete.*" (Sieden, 2011, p. 358).

Today we have the opportunity of a lifetime to innovate and inspire all of us to reimagine, redesign, reconcile, and regenerate our systems across all sectors so as to develop and transition to sustainability where all can thrive. And most importantly, education is at the forefront of these exciting new opportunities. In order to move forward effectively, we need a holistic, guiding educational framework that supports our transition to sustainability with regenerative socioecological relationships, socially just, resilient economies, and fair and equitable societies where individuals thrive and contribute to the health of our human and more-than-human communities on which we rely.

We are seeing the need to transition from an industrial societal worldview that saw humans as separate and superior to Nature to an ecological worldview that supports sustainability, seeing humans as part of Nature: influencing and being influenced by it. As humans are living systems, needing to develop systems that can adapt and emerge as conditions around us change, we can turn to the wisdom imbedded in Nature to mimic how it has evolved sustainable living systems.

Principles of Sustainable Living Systems

A living systems framework is based on the ecological principles all sustainable living systems depend on to be resilient: *interdependence, community, diversity, energy flows from the sun, cycling, feedback loops, adaptation, emergence,* and *the whole is greater than the sum of its parts* so we learn to thrive as we grow. As individuals and as a society we instinctively recognize the benefits of these ecological principles seeing them as positive social attributes. Humans are a natural species, integrally related to our complex, biodiverse living Earth system. We know this but need to learn to adapt better,

DOI: 10.4324/9781003389590-6

to operate "naturally" as part of Nature rather than trying to maintain the illusion we are separate and can overpower Nature.

Using ecological principles as an organizing framework may not yet resonate with some people in the mainstream as they have been taught to see the environment as separate from themselves and separate from society and the economy. Unfortunately, early environmental communication pitted the environment against the economy, reinforcing the concept of *the economy, society, and environment being separate*, effectively polarizing people into identifying with one or the other (Dale, 2002).

Yet we know all three imperatives are fully interrelated: the environment is our life-support system we are an integral part of, reliant as we are on clean air, water, food, materials, and living systems for our health and well-being. We have seen the rise of significant environmental issues due to our mechanistic worldview and are now recognizing how interdependent our environment, society, and economy are. Climate change is affecting our societies and economies with more severe and frequent extreme weather events; air and water pollution are directly impacting our health, food, and water. The United Nations Sustainable Development Goals (SDGs) (UN, 2015), developed by 150 countries around the world, recognize how interrelated social, economic, and environmental imperatives are, emphasizing each goal is affected by the others. This is supported by Richardson et al.'s (2023) research on the 9 planetary boundaries that shows how interdependent all sectors are with changes in one affecting the others, often in unpredictable ways.

We are beginning to recognize the wisdom of reinforcing the principles of sustainable living systems as we start transitioning to renewable energies and from a linear, take-make-waste economy, to a circular economy (Ellen MacArthur Foundation, 2017); in realizing our social health and well-being is improved by strengthening *community, interdependence* and *diversity*; and in finding our emotional and cognitive health is improved by spending time in Nature (Chawla & Escalante, 2007). These ecological principles are the foundation for thriving, so they provide the perfect framework for redesigning education and developing sustainability as a frame of mind (Bonnett, 2002), necessary for a sustainable future where all can thrive.

Redesigning Education

Our present school system has been useful in many ways and there are many examples of wonderful things happening in schools – but as we've seen, the system itself is not designed to support, and actually undermines necessary innovations through the influential root metaphors of the industrial mechanistic worldview (Ireland, 2007). We used to think we could maintain stability of our world through *efficiency, control, constancy, and predictability*;

attributes at the core of fail-safe design and optimal performance (Holling et al., 2002), hence these various attributes are embedded in our educational system (Ireland, 2007).

But natural systems, including growing, learning, human beings, are constantly adapting, changing, and emerging – that is, learning as conditions around them change as they develop and grow. In such dynamic, emerging living systems, stability, growth, and learning are not maintained through *efficiency, control, constancy, and predictability* but through *persistence, adaptiveness, variability, and unpredictability*. Our world is significantly changing, and with issues of climate change, global economic instability, and social injustice we need to fundamentally rethink and redesign education to be able to respond effectively.

Ministries of Education are trying to respond to many these challenges, but as we have seen, new approaches are typically grafted on a problematic mechanistic foundation that does not support innovations that are needed. There have been numerous innovations in administration, curriculum, and teaching over the years, but as we've seen, these have been piecemeal, subverted by the mechanistic system, thereby ineffective in changing the status quo. Educators, parents, and students understand we are living in an interconnected world, but few have realized our well-intentioned educational improvements are being constrained and undermined by our ingrained mechanistic foundations. We cannot continue to educate people in a linear, outdated system that is in direct conflict with societal needs. In recognizing we have been constrained by the old industrial, mechanistic worldview, we now have an opportunity to respond with insights gained from the interdependent systemic reality of the world, which functions as we do, as nonlinear living systems rather than as a dysfunctional machine. School needs to prepare us to develop our creativity and the systems thinking needed in our complex interconnected reality, one that develops a socially just, resilient economy that works within and enhances, rather than degrades, the natural environment we depend on for our life-support systems. Until we step back and consider the mindset we are schooling children into, and change that, we will continue to promote an outdated status quo mindset. What would a school that is good for students, good for society, good for the economy, and good for the Earth look like?

Simply bolting new curricula onto an outdated industrial-age school system won't work. As noted, we need to build a new way forward, one that helps students think in systems and relationships rather than parts; and develop resilience in complex emerging realities through persistence, adaptiveness, variability, and responding well to, even embracing unpredictability. In redesigning education, it is important to look at schools as complex living systems, where *interdependence* and *feedback loops* connect learning, students, teachers, administration, communities, and the natural world. This means not only redesigning how and where we teach, but also how we organize curricula, and how we structure and administer education.

We need a new model for education, one that helps the whole child learn, one that helps us understand how interconnected we all are and how we can adapt better to our fast-changing world. The good news is that model has already been designed for us. Ancient wisdom and cutting-edge research are now coalescing on the principles of natural living systems, developed over the past 3.8 billion years. We have realized these ecological principles are the key to thriving. Illuminating and applying these insights can help us redesign how our human systems can adapt to interact effectively with our natural life-support systems.

By focusing on ecological principles of sustainable living systems as our guiding framework we effectively replace the unsustainable root metaphors of the outdated industrial system with sustainable eco-centric root metaphors. As Lakoff (2010) emphasizes, metaphors influence how we think and act. To make systemic changes, we need a coherent framework that can guide transformation across the system such that all aspects and levels in the educational system are consistent with sustainability, leading to what Sterling (2001) refers to as sustainable education. Princen (2010) notes that by speaking differently through metaphors, we can inspire change and shift our worldview. In a sustainable world, Princen (2010) explains that metaphors are adaptive and have an ecological-human connection. He argues that metaphors are more than words; they are concepts and ideas that "Connect how we perceive with what we believe" and "Establish a worldview that guides ... how we relate to our environment" (p. 61).

We know, from our explorations in Part I, that influential and counterproductive mechanistic root metaphors from our industrial past are unwittingly incorporated into and undermining progress to bring about the necessary changes in education from administration, buildings, grounds, and resources to curriculum, teaching, and learning. This knowledge helps us understand how education and its foundational root metaphors need to fundamentally change if we are to help students develop the competencies needed to live and work sustainably in the complex world, we now find ourselves immersed in.

Sterling (2017, p. 1) underlines the importance of redesigning education based on an ecological worldview:

> A collective blindness to the global systemic issues that are shaping the near human and planetary future is present both in wider society and in educational systems that can, consequently, be deemed maladaptive to this reality. A deep learning response within educational thinking, policymaking, and practice is required based upon an emerging relational or ecological worldview, already burgeoning in diverse civil society movements.

As the ecological principles enable sustainable living systems to function effectively and be resilient, they provide the simple, elegant framework we are looking for and inspire effective eco-centric metaphors to replace problematic mechanistic ones. As Benyus (2014) says, to learn how to adapt to be sustainable we only need to look to Nature as our model, mentor, and measure. Taking a holistic, systemic approach to redesigning education and aligning the structure and functioning of education and schools based on ecological principles of sustainable living systems will help us transition from the problematic industrial worldview to a sustainable eco-centric or ecological worldview. In this way, innovations and individual efforts on the part of administrators and teachers will no longer be isolated, subverted by the "hidden" curriculum, and its powerful, dominant metaphors (Ireland, 2007).

So here we find ourselves – at an exciting turning point and opportunity for education. In responding to the realities of our interconnected, rapidly changing world in the 21st century, our educational structures and practices have the potential to help all students thrive; preparing us to engage effectively in our world; to live sustainably integrating our society and economy with our natural environment.

Education founded on the same principles of sustainable living systems can enable society to be responsive to the pressing issues and dilemmas of a changing world as it responds to worldwide economic pressures, social injustice and makes the necessary shift to a low-carbon sustainable society. It is the means to help students of all ages become literate and capable of taking their place in developing a sustainable society through their careers and life choices. In contributing to this exciting period of innovation, we have an opportunity to align education with sustainable living systems, so it can effectively adapt and no longer continue to transmit an outdated, unsustainable, industrial mindset. Let's imagine … what could schools that did this look like? If we focus on the design principles of sustainable living systems, to create a sustainable living educational system we will see improvements in how we organize and administer education, how we design the curriculum, how we teach as well as where and how we learn. But first, it's important to understand a living systems perspective, and the guiding ecological principles it is based on.

A New Way Forward: A Living Systems Perspective

A living systems perspective conceptualizes an educational system that learns from and applies ecological principles through its very structure. Rather than being a linear, machine-like system based on a top-down dominator model, it looks more like a living, dynamic, complex, self-regulating system of *feedback* loops and iterative thinking. If we look at the findings of science in the 20th and 21st centuries, we see the natural world is not made up of separate, independent particles as Descartes and Newton suggested. Rather, science has found that in living systems *the whole is greater than the sum of its parts*, so the focus is on *communities* rather than individuals, seeking to optimize systems rather than maximize components.

Rather than following a linear industrial model, sustainable education, that is, education for sustainable development is based on this nonlinear perspective. In this holistic nested model, the individual is viewed in context as an integrated, *interdependent* aspect of the whole of society and the natural environment. The whole of the student, their mind/body/environment, is seen in relationship rather than as separate. This parallels the nested reality of living systems where organisms are integrally *interdependent*, influencing as well as being influenced by their *communities* and ecosystems.

Beyond seeing the individual systemically, nested in larger systems, sustainable education follows a similar nested philosophy where education serves to develop and enhance the *community* and a sustainable society that functions within the natural environment. Such living systems or *ecospheric* models are dynamic, *interdependent* systems with sensitive *feedback loops* between all levels in the system from environment to society, community, education, and students who are dynamically interacting, *adapting*, and *evolving*, as seen in Figure 3.1.

This nonlinear living systems perspective incorporates *interdependence* within and between systems as dynamic *community* relationships; honours

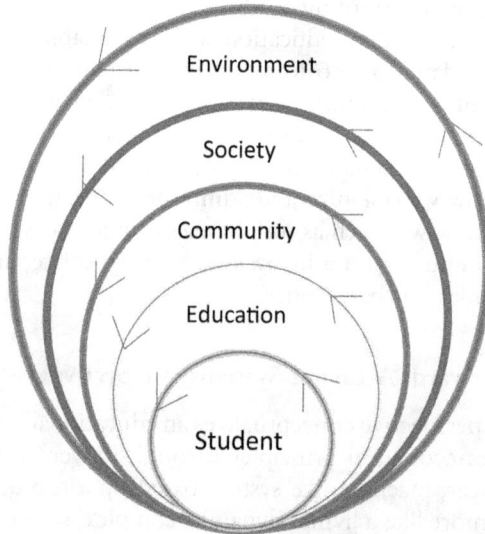

Figure 3.1 Nested interdependent educational system. Created by the author.

and celebrates *diversity* and development as increasing complexity within a systemic context; and iterative *cycling* with *emergent* properties unfolding and enfolding providing constant *feedback, adaptation,* and change. This systemic worldview encourages an emergent, transformative, integrative view of knowledge and learning and can be depicted as shown below by a tree metaphor for education in Figure 3.2.

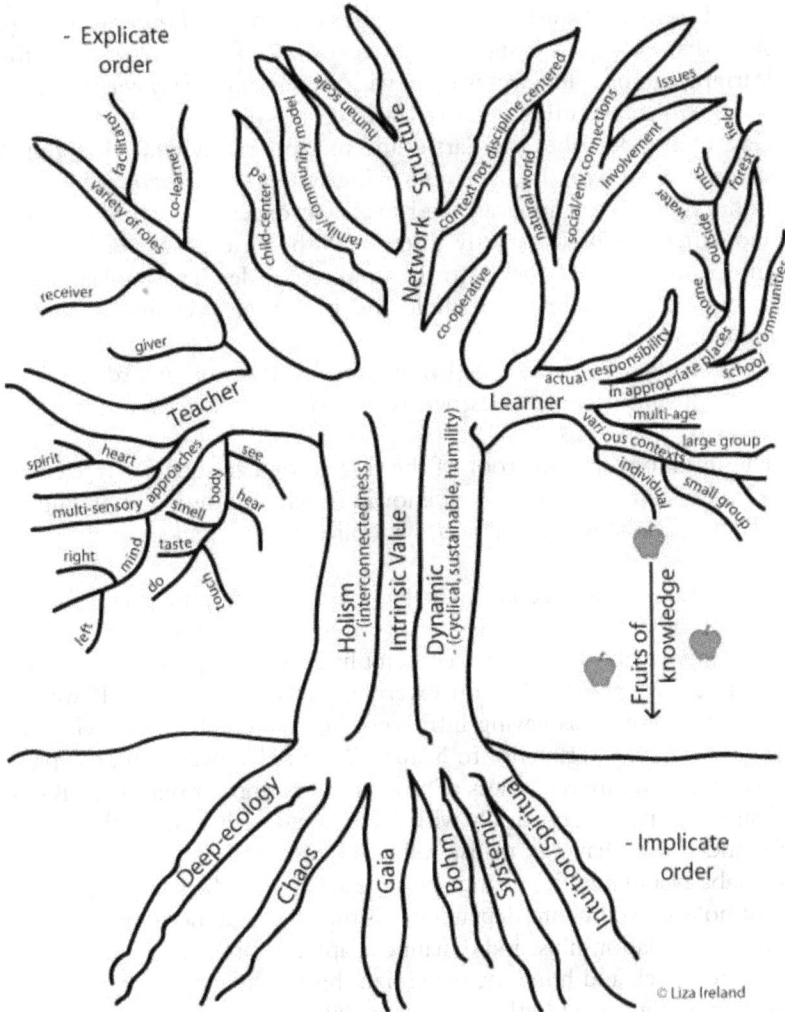

Figure 3.2 Education as a complex living system. Created by the author.

As a complex living system, the roots of education support and nourish an emergent understanding of students as integral parts of living systems:

- A deep ecological understanding recognizes our socio-ecological *interdependence*, as well as resilience and thriving depend on the integrity of the whole system where learning occurs from the student through to the Ministry of Education, the society, and environments they are immersed in.
- Chaos theory shows that through a flow of energy in relation to changing conditions, order (or understanding of increasing complexity) emerges out of seeming disorder; new structures and forms of order can *emerge*, such that change and stability can co-exist (Capra, 1996). Therefore, learning can be stable, yet is open with a constant *flow of energy*, leading to unpredictable future paths of learning.
- Gaia theory describes the Earth (and metaphorically education and the student) as an *adaptive, emerging* living system with *intrinsic value*.
- Bohm's theory of implicate and explicate orders describes how inner and outer worlds constantly interact, with the explicate order arising and being enfolded back into the implicate order. The implicate order is implicit, underlying the whole and is invisible, yet fundamental, to understanding (Bohm, 1985).
- Systems theory recognizes living systems function interdependently at various interacting scales, where *the whole is greater than the sum of its parts*.
- The intuition/spiritual root of the tree recognizes qualitative elements of reality and multiple ways of knowing such as intuitive, tacit, cultural, affective, aesthetic appreciation, symbolic, creative, and artistic.

The trunk of the tree is supported by three key principles. The core recognizes and honours life as having *intrinsic* value, thereby incorporating the importance of *diversity*. This emphasizes the importance of moving beyond valuing based solely on external, instrumental value. If we view Nature and humans as having intrinsic value, we simply cannot view ourselves as separate and superior to Nature, dominating over it and having the power to use Nature regardless of repercussions to meet our goals. Rather, we learn and live synergistically with Nature and each other with both humans and Nature having intrinsic value. Holistic and dynamic concepts of sustainability and iterative change provide stability to the trunk. The principles of holism, where interdependence is manifest and meaning arises from context and relationships; and dynamic adaptation and emergence through cyclical feedback and humility, recognize this systemic concept of learning leading to resilience, growth, and sustainability.

In later chapters we will explore how these principles provide a framework to support effective transformations in the branches of the tree through administration, curriculum, buildings/grounds, teaching, and learning. Before we get there, it's important to look a little more closely at how complex living systems thrive through the core ecological principles, as these inform and create the conditions conducive to thriving, learning, and growing (Robinson, 2016).

Interdependence

Within an ecosystem, everything is interconnected and interdependent. In forests, trees are connected through an interactive web of mycelium that helps trees communicate (Wohlleben, 2016). Bees or birds pollinate the tree's flowers, while the flowers and fruit provide them with food. They share resources and waste from one species becomes food for another. Red squirrels in boreal forests, for example, feed on an underground fungus that helps trees absorb water and nutrients. Trees provide safe habitats for squirrels and squirrels help new tree seedlings grow by spreading fungal spores in their droppings over the forest floor. Species are interdependently related to each other and the systems that support them.

In education these interconnections need to become more apparent. Seeing the various aspects of the educational system as an intricate web will help. Challenging discipline barriers is a good first step to seeing how interconnected things really are. It is starting to become more common for Ministries of Education to support teaching and learning through topics rather than subject disciplines to help students understand interconnections. This way students and teachers learn about topics such as climate change by incorporating diverse perspectives, see how all the various subjects interdependently inform that topic. To understand complex systems such as climate, students need to explore how the climate system works from a scientific point of view as well as the political, economic, and social issues that are related to climate change. Through this integrated exploration of a topic, math, language, and artistic skills are needed to develop and express the students' learning. This approach would help students engage with and understand complexity. The emphasis shifts to systems and interconnections, helping students to think in systems, rather than focusing on separate parts.

Community

Within a forest ecosystem, organisms create communities in order to live together, making the best use of their resources be they considered food or waste. Bees, trees, squirrels, worms, bacteria, and larger or smaller plants and animals all work together to create a healthy forest ecosystem and the conditions conducive to life.

In applying a living systems perspective to education, students, educational staff, and communities rise to their full potential working together to help create a sustainable, just, fair community for all: human and more-than-human. Rather than focusing on individuals being responsible for solving the many issues we face, it will be more about "us," working and learning in reciprocal, interdependent networks to realize greater potentials in community.

Diversity

In a forest, diversity also plays an essential role as it ensures a store of potential innovation and adaptive responses are available as conditions change. A failure to release this creativity for the next phase creates a rigid, fragile system no longer able to adapt as conditions change. It is not surprising then, that diversity is inherent in Nature as it provides strength and resilience. In society we recognize the importance of honouring diversity.

It's important to consider how we can infuse and support diversity as a core construct of our system of educating to maintain resilience. We need to consider how we can recognize and empower diversity in and through administration, curriculum, teaching, and learning.

Energy Flows from the Sun

Living systems are powered with energy flows from renewable, current sunlight, ensuring a constant supply of energy. Energy also flows through the diverse species that make up an ecosystem.

The hidden curriculum is very powerful in supporting or subverting the overt curriculum. In aligning education with the principles of living systems, educational buildings, therefore, should be recognized as part of the curriculum, exemplifying ecological principles by being powered by renewable energies. Similar to energy flows in natural sustainable systems, energy from all stakeholders in the educational system needs to flow between all levels in the panarchy.

Cycling

Through these interconnections sustainable systems use resources effectively, relying on energy and nutrients being continually cycled. In forests, plants take up nutrients from the soil, animals eat these plants, larger animals may eat smaller animals, and then when the animal dies, it decomposes adding nutrients back to the soil. Living systems exemplify this complexity where waste from one source is food for another, and materials are continually cycled and reintegrated. Seeing and thinking in systems rather than distinct parts will help us understand interrelationships and how all things naturally cycle in sustainable systems.

In education we need to understand how things cycle in systems rather than in linear trajectories, as exemplified in our take-make-dump societal practices. Learning is also cyclical as we benefit from revisiting, re-thinking, and revising to extend and deepen our concepts and understanding.

Feedback

Through effective feedback loops, information helps a system adapt and self-correct, or reinforce, effective actions, which are an essential feature of resilient, sustainable systems. In this way, feedback allows natural systems to adapt as conditions around them change.

Similarly, to enable a sustainable educational system, we need to develop more effective processes of feedback to help empower change at all levels in the system. Feedback needs to be based on strong, effective interdependent relationships between students, teachers, parents, community, administrators, and maintenance departments. Students are typically not given many opportunities to influence their own education through feedback. Don't get me wrong – feedback happens – but it is often not welcome unless it's an "A" grade for the student or cooperative behaviour from the students. We're not really sure what to do with the students who find school is not working for them and changing how we educate based on individual needs isn't easily incorporated. Effective feedback would encourage innovation and allow learning to emerge from students' interests and local community needs, helping students own their learning and feel part of their communities. This allows schools to become learning organizations, encouraging creativity and innovation at all levels.

Adaptation and Emergence

In studying sustainable living systems, we've also learned that through feedback, change and adaptation are essential aspects that ensure a resilient system able to evolve as conditions change. The forest is never frozen in time. The whole forest and the various components adapt to changes in the seasons, patterns of growth, insect infestations, invasive species, population dynamics, and changes in the climate. This change happens through self-assembly and emergence, rather than planning and imposing solutions from the top down.

Holling and Gunderson (2002) developed an elegant model to show how complex adaptive systems continually evolve through an adaptive cycle, as shown in Figure 3.3.

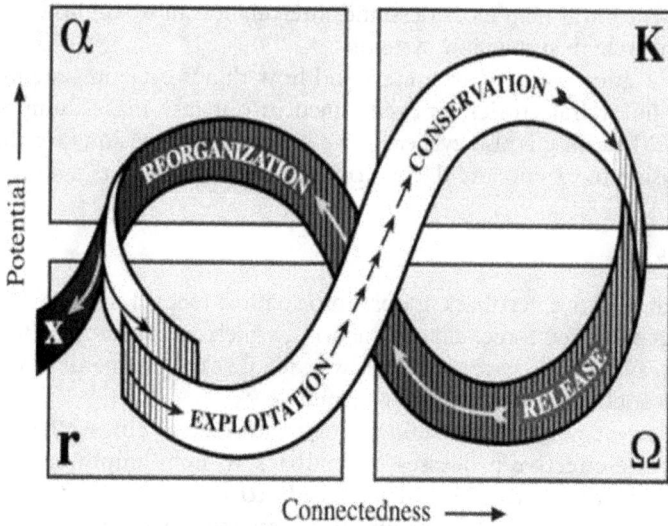

Figure 3.3 From *Panarchy* edited by Lance H. Gunderson and C.S. Holling., Chapter 2, Figure 2-1. Copyright © 2002 Island Press. Reproduced by permission of Island Press, Washington, DC.

All ecosystems, from the cellular to the global level, are said to go through these four stages of a dynamic adaptive cycle.

- The *exploitation* stage (r) is one of rapid expansion, as when a population finds a fertile niche in which to grow.
- The *conservation* stage (K) is one in which slow accumulation and storage of energy and material is emphasized as when a population reaches carrying capacity and stabilizes for a time.
- The *release* (Ω) occurs rapidly, as when a population declines due to a competitor, or changed conditions.
- *Reorganization* (α) can also occur rapidly, as when certain members of the population are selected for their ability to survive despite the competitor or changed conditions that triggered the release (The Sustainable Scale Project, 2006, para 5).

For example, on the west coast of Canada, the coastal ecosystem generates excellent conditions for Douglas fir trees to establish themselves; overtime a mature forest and old growth trees become established; in the event of a fire, release occurs, causing a renewal and releasing nutrients (K) that allow for new growth. This move into the back loop brings forth opportunities and innovation, increasing potential in the next phase of development. In the new stages of exploitation and conservation, potential and resilience are increased as interconnections are re-established, and dominant species establish themselves, until another adaptive cycle allows new innovations to emerge as conditions change. This provides both strength and resilience.

A failure to release for the next phase, as conditions change, creates a rigidity trap, preventing release and reorganization, and loss of resilience. This seems to be where our present educational system is stuck, trying to maintain the status quo. While we tend to stress competition and *survival of the fittest*, Darwin's contributions to our scientific understanding also emphasized it is *those species best able to adapt that survive*. This is an inspiring framework for education as a resilient, sustainable system is based on continuous learning and adaptation: releasing and letting go of old patterns or associations as conditions change and new ideas present themselves; engaging innovation and creativity to explore new ideas and reorganize our thoughts; determining new ways forward; and then establishing new understandings, skills, and patterns of behaviours.

In natural systems adaptation happens at all levels in the system in a dynamic dance of change and adaptation. Change is not a struggle between grassroots innovation or control from above. Adapting and evolving are both. To ensure sustainability, natural systems are part of nested hierarchies that dynamically interact, in what Holling et al. (2002) term a "Panarchy," as shown in Figure 3.4.

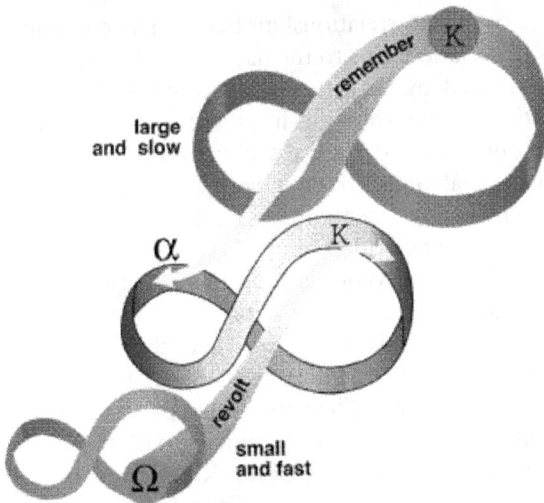

Figure 3.4 From *Panarchy* edited by Lance H. Gunderson and C.S. Holling., Chapter 3, Figure 3-10. Copyright © 2002 Island Press. Reproduced by permission of Island Press, Washington, DC.

As adaptive cycles are happening at all levels in the system, change at a smaller faster scale can influence or cascade up to a larger, slower scales by triggering adaptations in that higher level. Conversely, the larger, slower level stabilizes by remembering successful innovations. So the fast levels invent, experiment, and test, while the slower, larger levels stabilize and conserve from accumulated memory of successful past experiments. This combines learning with continuity to create a stable yet dynamic evolving system. Unlike our dominant school system, diversity, innovation, creativity, adapting, and remembering are built right into natural systems (Holling et al., 2002).

The Whole Is Greater than the Sum of Its Parts

All this shows that one of the greatest principles of living systems is *the whole is greater than the sum of its parts*. In living systems all the various parts interact such that the inter-relationships give rise to a synergy such that the system takes on an energy and function beyond what each part contributes on its own. As such, the system has unique properties that arise through collaboration. Yet each part is essential as it contributes to the dynamic health of the whole. The forest is more than a collection of animals and plants, rocks, air, and water. All the diverse living and non-living parts of the forest interact together, creating dynamic systems that continually adapt and evolve, influencing every aspect of the forest. As well, the forest as a whole acts to moderate larger systems it is a part of, such as climate.

Understanding this as a model in education is essential. It strengthens the importance of the interrelationships between all components in the system as well as the relationships to the natural and human systems education affects and is affected by. This ecological principle highlights the importance of holistic, systems thinking in administration; buildings, grounds, and resources; curriculum design; teaching and learning. It is essential for the whole educational system to embrace an ecological worldview as working together, reinforcing interconnections between all parts, at all levels, can lead to systemic transformation.

We can exemplify the ecological principles of sustainable living systems through the various components of a forest and natural ecosystem, but as humans are also part of this rich web of life, this model is not complete until we put people into the picture. It is important to emphasize the significance of relationships in context and seeing patterns that connect people to people – historically and intergenerationally, with those who think differently, or may have Traditional Ecological Knowledge; through horizontal and vertical relationships in systems – within and between the school and community; and through our socio-ecological relationships. Understanding we are part of interconnected, continually adapting systems will help us develop an exciting sustainable future full of new opportunities, and an educational system that can respond to the changing needs of society while creating the conditions for students to thrive.

Conclusion

By looking at how sustainable, interdependent living systems function we can redesign education to develop not only a mindset and way of thinking that is in line with, rather than counter to Nature and our critical support systems, but also one that is engaging and relevant to help us transform ourselves and our society to be more adaptive and resilient. If the role of education is to help all students learn, to develop and realize their full potential as contributing members of a sustainable society, to help transition to a low-carbon sustainable society, and prepare students to continue this process through critical, creative thinking using multiple intelligences, considering multiple perspectives – carrying this through their careers and life choices – then the system of schooling as well as its infrastructure and methods will also need to transition to support this purpose.

An eco-centric framework will help us do just that. As we face the challenges before us in rethinking education in the 21st century, we have an exciting opportunity to use these principles to help us redesign education so as to avoid a rigidity trap that hinders or undermines adaptation and learning. Many of these principles are already recognized as diversity, innovation, creativity, adapting, and continuous learning are the hallmarks of 21st-century educational innovations. But we also need to look at how can we design an educational system that supports rather than subverts these innovations; and supports transformative education we envision for students to realize their full potentials. Ultimately, we need to consider how we can use these principles to create the mindset needed to develop a socially, environmentally, and economically sustainable society that functions in concert with its natural foundation rather than in opposition to it.

An eco-centric framework based on the ecological principles of sustainable living systems unites educational initiatives together into a solid philosophical framework that helps us respond effectively to the challenges we face in education and society. Throughout the educational system, all have opportunities to be part of rethinking and redesigning education. Let's now look more specifically at how administration, buildings and grounds, curriculum, teaching, and learning transform when they are based on ecological principles. We all care deeply about providing our children with the best education we can. It's time to create the conditions so all thrive.

Note

1 Sieden, L. S. (2011). A Fuller View – Buckminster Fuller's Vision of Hope and Abundance for all. *Divine Arts Media*, p. 358.

References

Benyus, J. (2014). Biomimicry in action. *TED Talk.* https://www.ted.com/talks/janine_benyus_biomimicry_in_action?language=en

Bohm, D. (1985). *Unfolding meaning: A weekend of dialogue with David Bohm.* Ark paperbacks.

Bonnett, M. (2002). Education for sustainability as a frame of mind. *Environmental Education Research, 8*(1), 9–20. https://rrojasdatabank.info/sustbonnett.pdf

Capra, F. (1996). *The web of life: A new synthesis of mind and matter.* Flamingo.

Chawla, L., & Escalante, M. (2007). *Student gains from place-based education.* Children, Youth and Environments Center for Research and Design, University of Colorado. Retrieved from https://promiseofplace.org/research-evaluation/research-and-evaluation/student-gains-from-place-based-education

Dale, A. (2002). *At the edge: Sustainable development in the 21st century.* UBC Press.

Ellen MacArthur Foundation. (2017). *The circular economy.* Retrieved from https://www.ellenmacarthurfoundation.org

Holling, C. S., & Gunderson, L. H. (2002). Chapter 2: Resilience and adaptive cycles. In L. H. Gunderson & C. S. Holling (Eds.), *Panarchy: Understanding transformations in human and natural systems.* Island Press.

Holling, C. S., Gunderson, L. H., & Peterson, G. D. (2002). Chapter 3: Sustainability and panarchies. In L. H. Gunderson & C. S. Holling (Eds.), *Panarchy: Understanding transformations in human and natural systems.* Island Press.

Ireland, L. (2007). *Educating for the 21st century: Advancing an ecologically sustainable society* [Doctoral dissertation, University of Stirling]. STORRE. http://hdl.handle.net/1893/240

Lakoff, G. (2010). Why it matters how we frame the environment. *Environmental Communication, 4*(1), 70–81.

Princen, T. (2010). Speaking of sustainability: The potential of metaphor. *Sustainability: Science, Practice and Policy, 6*(2), 60–65. https://doi.org/10.1080/15487733.2010.11908050

Richardson, K., Steffen W., Lucht, W., Bendtsen, J., Cornell, S. E., Donges, J. F., Druke, M., Fetzer, I., Bala, G., Von Bloh, W., Fuelner, G., Fiedler, S., Gerten, D., Gleeson, T., Hofmann, M., Huiskamp, W., Kummu, M., Mohan, C., Nogués-Bravo, D, … Rockström, J. (2023). Earth beyond six of nine Planetary Boundaries. *Science Advances, 9*(37). Retrieved from https://www.science.org/doi/10.1126/sciadv.adh2458

Robinson, K. (2016). *Sir Ken Robinson & Dr. Peter Senge – Education fit for the 21st century.* Disruptive Innovation Festival. Retrieved from https://www.youtube.com/watch?v=j1egRlszeH4

Rockström, J. W. (2015). *Bounding the planetary future: Why we need a great transition.* Tellus Institute. Retrieved 1 June 2016, from http://www.tellus.org/pub/Rockstrom-Bounding_the_Planetary_Future.pdf

Sieden, L. S. (2011). A Fuller view – Buckminster Fuller's vision of hope and abundance for all. *Divine Arts Media,* p. 358.

Sterling, S. (2001). *Sustainable education: Re-visioning learning and change.* Green Books.

Sterling, S. (2017). Assuming the future: Repurposing education in a volatile age. In B. Jickling & S. Sterling (Eds.), *Post-sustainability and environmental education. Palgrave studies in education and the environment.* Palgrave Macmillan. https://doi.org/10.1007/978-3-319-51322-5_3

The Sustainable Scale Project. (2006). Retrieved from http://www.sustainablescale.org

UN. (2015). *Sustainable development goals.* Retrieved from https://sustainable development.un.org/?menu=1300

Wohlleben, P. (2016). *The hidden life of trees: What they feel, how they communicate.* Greystone Books Ltd. David Suzuki Foundation.

4 Organization, Administration, and Leadership to Support Transformative Learning and Systems Thinking

How we structure and administer education is an indication of our worldview and has a profound impact on how education is experienced by all (Ireland, 2007). If we are to transform education based on the ecological principles of sustainable living systems, to encourage systems thinking and the ability to adapt and innovate, how we structure and administer education needs to transform. If not addressed, educational administration and leadership can undermine transformative initiatives and continue to reinforce the unsustainable status quo explored in earlier chapters. No matter how innovative a teacher wants to be, if education is being controlled and administered to a predetermined set of mechanistic root metaphors, creativity and transformative education will be undermined, as teaching and learning are narrowly directed towards predetermined outcomes within limiting structures.

If it is given careful consideration, how we organize and administer education can transition to enable a system of education that can further the development of a learning society and sustainable worldview. Several organizational and business authors concur having proposed an updated conceptualization of organizations as organic systems, or "living organizations" (Vlados, 2019, p. 230) rather than mechanistic ones. In discussing change management, Vlados (2019) calls for "The gradual transition from the analytical perspective of classical/mechanistic management to the "organic" management" (p. 234), saying, "...organizations are not machines but complex adaptive systems within an evolving environment" (p. 241). The ecological principles of sustainable living systems provide the framework to guide this transition. Looking at each principle in turn helps us see how each principle informs effective organization, administration, and leadership, and how together, they interact to develop a holistic, dynamic, responsive system to support and develop an educational mindset and practice based on creativity, systems thinking, diversity, decolonization, and a sustainable worldview.

DOI: 10.4324/9781003389590-7

Ecological Principles Informing Organization, Administration and Leadership

Interdependence

As we saw in the last chapter, sustainable living systems are resilient as they create the conditions conducive to thriving, learning, and growing through a nested, interactive, interdependent panarchy, exemplified in Figure 4.1.

As described in Chapter 3, a panarchy is a nested hierarchical structure in which systems of Nature, as well as socio-ecological systems, are interlinked in never-ending adaptive cycles of growth, accumulation, restructuring, and renewal (Holling et al., 2002). Unlike traditional hierarchies, each level in the panarchy has the ability to adapt to changing conditions, thereby developing variety and generating new experiments to learn from. Secondly, there are multiple connections between levels to create and sustain adaptive capability. A critical change at smaller, faster adaptive levels can cascade up to influence and cause change at a vulnerable stage in a larger, slower level causing what Holling et al. (2002) refer to as "revolt." Yet, panarchical systems are also stabilized at times of change as levels are also connected from larger to smaller scales through a "remember" connection that facilitates renewal by drawing on the potential that has been accumulated and stored in the larger, slower cycle.

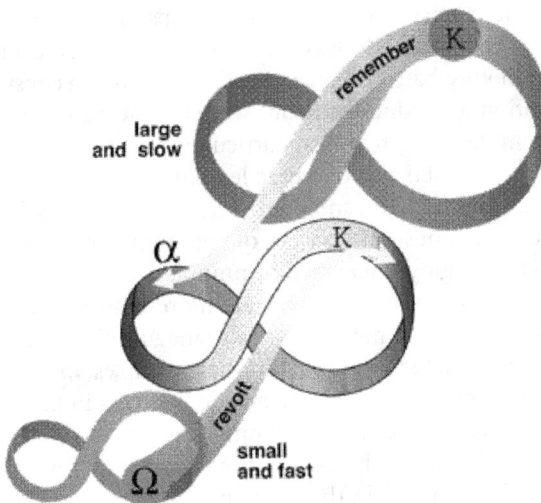

Figure 4.1 From *Panarchy* edited by Lance H. Gunderson and C.S. Holling, Chapter 3, Figure 3-10. Copyright © 2002 Island Press. Reproduced by permission of Island Press, Washington, DC.

This organizational model is the foundation of sustainable, resilient, living systems. As such, it is the critical organizational structure we need to enable sustainable learning systems in redesigning for sustainable education. As Holling et al. (2002) explain,

> [A panarchy] summarizes succinctly the heart of what we define as sustainability. The fast levels invest, experiment, and test; the slower levels stabilize and conserve accumulated memory of past successful, surviving experiments. The whole panarchy is both creative and conserving. The interactions between cycles in a panarchy combine learning with continuity. (p. 76)

This shows all levels in the panarchy are necessarily *interdependent*: the larger system stabilizes, maintaining innovations that create the conditions to thrive; while receiving constant *feedback* from the other interdependent levels in the system so as to embrace *adaptation* and *emergence* as conditions at various levels in the system respond and change (Holling et al., 2002). Applied to educational governance, this panarchic model creates an organizational structure that models and teaches an eco-centric, sustainable worldview, thereby enabling sustainable education.

Community

In supporting multi-stakeholder, multi-scale transformative innovations, this sustainable living systems model is contextually responsive, creating a dynamic, interactive community-based organizational model. When considering educational organization and administration, we have the basic components of a school ecosystem: learners, teachers, curriculum, administrators, parents and culture as well as the influence of where learning happens: the buildings and grounds, human communities, and our natural environments. This ecosystem conceptualization expands our concept of community beyond the school by enabling and encouraging interactive communication up to the scales above in the panarchy, and out to our cultural and more-than-human communities. Westley (2002) describes a similar model as managing *through, in, out,* and *up*: *through* scientific understandings of ecological principles, *in* by being inclusive with those in your organization, *out* to include the community, and *up* to consider the levels above. The process of communicating with the community, in, out, up and through is not to be underestimated in a healthy, resilient system.
This is exemplified below in the "Systemic Effects" section.

Diversity

An organizational model that mimics organic, living systems necessarily incorporates diversity and pluriversalism. In acknowledging how important spatial-temporal context and other ways of knowing are, "pluriversalism" has recently risen to the forefront. To increase awareness about the concept of pluriversalism, the UNESCO Institute for Lifelong Learning (UIL) and the Sustainable Futures Global Network hosted a webinar in March 2023, entitled "Pluriversal Literacies for Sustainable and Equitable Education," wherein education stakeholders from policy, practice, research, and academia discussed how pluriversalism can be integrated into the way we think about, relate to, design, and create our future world (UNESCO Institute for Lifelong Learning [UIL], 2023). This recent initiative of UIL shows a growing interest in moving past the Western industrial educational paradigm, recognizing we need to create a new organizational structure and administrative system that incorporates multi-stakeholders and multiple ways of knowing. This important as it aligns well with our multicultural world and the need to decolonize education.

In decolonizing the Western education model, steeped in the mechanistic paradigm that has been exported around the world, it is important to reconsider the Cartesian concept of universality imbedded in applying this decontextualized mechanistic management structure. Grosfoguel (2012) speaks of Western universalism, derived from Descartes' abstract universalism as, "that of a universalism based in a knowledge with pretensions of spatio-temporal universality" (p. 90). As pluriversal thinking embraces and is grounded in spatial-temporal contexts, it gives voice to diverse ways of knowing and organizing. Therefore, in our efforts to recognize diverse socio-ecological relationships and decolonize education, pluriversalism is extremely important. As Pashby et al. (2020) note, "Thinking with pluriversality provides the possibilities for a praxis that works towards a re-existence of relationships across differences" (p. 47) and " A praxis of pluriversalism takes us beyond, while at the same time undoing, the singularity and linearity of the West" (Walsh & Mignolo, 2018, p. 3 as cited in Pashby et al., 2020, p. 48).

Developing a panarchical educational system highlights the importance of diversity through polycentric governance that recognizes multiple governing actors to guide interactions across similar scales horizontally and within nested levels vertically. According to Schoon et al. (2015),

> [Polycentric governance systems] have been found to create a foundation for learning and experimentation, to be a source of policy/institutional diversity, to enable broader levels of participation and to improve connectivity between groups while building in modularity and redundancy. (p. 226)

Decolonizing education through polycentric governance is being exemplified in creating a new school partnership between Snuneymuxw First

Nation and Nanaimo Ladysmith Public Schools British Columbia. According to Nanaimo Ladysmith Public Schools (2023), they are committed to co-govern and operate the new Qwam Qwum Stuwixwulh School. This is part of an ongoing journey to restore a sense of place and to connect students to the knowledge and language that is rooted in these sacred lands, teaching children to walk in "two worlds".

At each nested level in the educational system, groups of people self-order their relationships for deliberation and decision-making; and guiding principles help establish rules of interaction across a variety of scales to address problems or challenges confronted at different temporal and spatial scales due to systems dynamics and feedbacks occurring at these multiple scales (Schoon et al., 2015).

In this way, polycentricity attempts to match governance and administration to the scale of the problem and recognizes pluriversalism as individuals are encouraged to participate, sharing their knowledge and insight through social engagement and collective action, in decisions that affect them directly. As conditions in complex adaptive systems (CAS) are constantly needing to respond to changing conditions at various scales, polycentric systems of governance, rather than top-down centralized control, allow for experimentation across multiple authorities at a given level as well as learning across them (Schoon et al., 2015, p. 228). This enables an educational system that is designed for learning; one that is adaptive, more resilient, inclusive of diverse multi-stakeholders, including First Nation elders and knowledge keepers in locally appropriate decision-making at a variety of levels.

In this polycentric model, changes are enabled from all levels: students, teachers, and parents are involved in teaching and learning decisions at the class level; students, parents, and community members, as well as teachers and the Principal, are part of school-based planning; students, parents, and teachers have an important voice through having a formal position on the school board in district level decision-making; and they also form a consultative group with representation across districts for discussions, particularly as they relate to curriculum, at the ministerial level. School administrators, teachers, students, parents, and community members have specific roles and unique perspectives, yet they are empowered decision-makers, working in consultation with each other.

Cycling and Feedback

In sustainable living systems feedback and change come from all levels in the system because of multi-direction, cyclical feedback loops in the panarchy. Multi-stakeholder committees with established feedback across and between levels in the system are integral to the system design. Feedback loops within levels and between all levels in the system help evolve appropriate, effective solutions. Through effective channels of feedback, faster, grassroot innovations in teaching and learning can influence teaching, curricula as well as administration while slower higher level remembrance of successful implementations can help stabilize and guide – as long as all levels remain open to effective communication and innovation (Figure 4.1).

Currently many Ministries and Boards of Education are trying to strengthen feedback from all stakeholders, but these efforts are being grafted onto the traditional organizational structure (Figure 2.1) where decisions and changes are planned and imposed from the top-down with very limited feedback mechanisms between levels or from the multi-stakeholders. In supporting the panarchy, feedback strengthens interdependence and community, incorporates diversity through pluriversalism and polycentric governance. Reciprocal feedback between all stakeholders, students, teachers, support staff, buildings and grounds personnel, parents, and community members, is essential to create more effective communication, better learning, and a dynamic system that can adapt as conditions change.

Adaptation and Emergence

This interdependent, dynamic, community-based, model that incorporates diverse, pluriversal perspectives and constant feedback enables resilience through adaptation and complex adaptive systems (CAS) thinking. The importance of CAS thinking in developing resilience cannot be overstated. As Schoon et al. (2015) emphasize,

> CAS thinking is essential to create the awareness and mental models needed to inform new models of governance and management that can support these outcomes (managing for diversity and redundancy, connectivity, participation, and learning) and address key feedbacks of SES [socio-ecological systems]. (p. 259)

Developing a panarchical, polycentric system of governance applies CAS thinking in recognizing multiple governing actors to guide interactions across similar scales horizontally and within nested levels vertically. Yet as Schoon et al. (2015) note, all those in the educational system need to develop and embrace complex adaptive systems CAS thinking, seeing themselves as interdependent, contributing members of a responsive adapting community they influence and are influenced by. This underlines the importance of professional development for administrators and teachers, elaborated on further in the "Policy" section.

In transitioning from a centralized, top-down plan-and-impose organizational structure, control becomes decentralized, enabling internal, democratic, and adaptive decision-making and actions, as conditions change, and learning happens. As mentioned earlier, sustainable living systems are not stabilized through *efficiency, control, constancy, and predictability,* hallmarks of traditional top-down management systems, and the rationale for maintaining centralized control. Rather, stability is maintained through management and governance that embraces *unpredictability, change, adaptiveness, and persistence* (Holling & Gunderson, 2002): that is, stability is created through learning and adapting across all levels in the panarchy.

Looking more closely at adaptation, each level in the panarchy incorporates adaptive cycles. Holling and Gunderson (2002) have found that natural ecosystems as well as human systems move through four stages in an adaptive cycle in response to shifting conditions: release, reorganization, exploitation of a new opportunity, and conservation. The same can be said of our social systems when they are allowed to adapt to changing conditions as learning occurs (Figure 4.2).

As described in Chapter 3, the adaptive cycle shows a release from the status quo (K), an organizational pattern or way of thinking that had been conserved and maintained will quickly lead to reorganizing in the back loop in adapting to the new stimulus and changing conditions. New ideas will be exploited in the front loop and then slowly established and conserved as potential and connections increase. With increasing connectedness there will

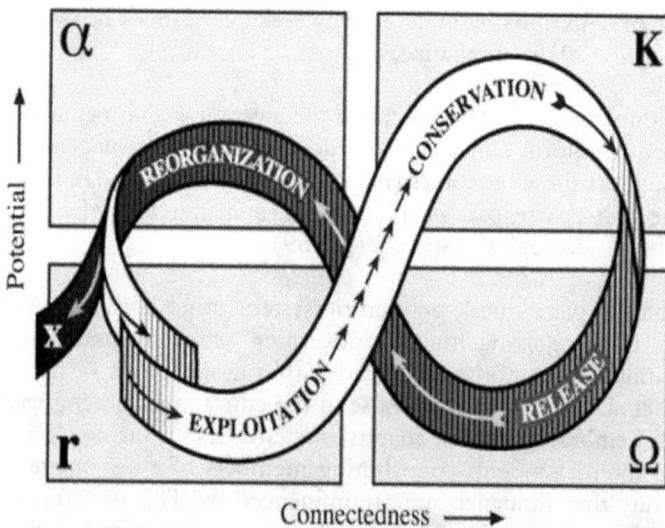

Figure 4.2 From *Panarchy* edited by Lance H. Gunderson and C.S. Holling, Chapter 2, Figure 2-1. Copyright © 2002 Island Press. Reproduced by permission of Island Press, Washington, DC.

eventually be less flexibility and resilience. But as the adaptive cycling enters a new cycle of release, reorganization, and exploitation, resilience, or the ability to adapt as conditions change, increases. So, the ability to continually adapt and change, that is, to learn as conditions change, is the key to being resilient or stable by embracing unpredictability, change, adaptiveness, and persistence.

We often see established bureaucratic systems, such as the industrial, mechanistic educational system becoming over-connected and increasingly rigid in its control. However, a stimulus or change in conditions, such as the realities of the 21st century and the need to transition to a sustainable society, can eventually initiate a release and reorganization of how we respond and educate. Reorganizing the educational system as a panarchy will have a cascading effect right through all levels in the system. At the smaller scales students and teachers will constantly learn through adaptive cycles, letting go of previous concepts or understandings in light of new feedback; reorganizing; and exploiting new ideas, theories, or strategies. New experiments and innovations can cascade up to higher levels influencing other classes at the school, district, or ministerial levels. Similarly, each level in the system has the ability to adapt as changing conditions at those levels indicate the need for further adaptations.

To help with this dynamic interplay within and between levels in the system, adaptive co-management is an increasingly popular and effective administrative approach when working with CAS. In contrast to the "command-and-control" structure, the adaptive co-management approach is flexible, working with organizations at different scales; it highlights group decision-making process, diverse knowledge and views, adaptability, resilience, and transformation (Armitage et al., 2009). Olsson et al. (2004) further note, "Adaptive co-management relies on the collaboration of a diverse set of stakeholders operating at different levels, often in networks, from local users, to municipalities, to regional and national organizations, and also to international bodies" (p. 75).

In adaptive co-management, the emphasis is on experiential and experimental learning in addition to vertical and horizontal collaboration across all scales necessary to improve our understanding of, and ability to respond to, complex social-ecological systems (Armitage et al., 2009). The development of adaptive co-management systems shows how local groups can self-organize, learn, and actively adapt to and shape change with social networks that connect institutions and organizations across levels and scales and that facilitate information flows (Olsson et al., 2004). This process helps support polycentric, pluriversalism discussed earlier.

This information flow needs to enable and support decentralized decision-making, "Novel governance approaches emphasize group decision making that accommodates diverse views, shared learning, and the social sources of adaptability, renewal, and transformation" (Armitage et al.,

2009 p. 96). This depends on effective linkages to enable regularized flows of information, shared understandings, and problem articulation to move governance beyond simplified network perspectives (Armitage et al., 2009). The authors go on to identify this learning model, which accounts for social context (including power imbalances), pluralism, critical reflection, adaptive capacity, systems thinking, or interconnectedness, a diversity of approaches to adaptation, and paradigm shifts is essential (p. 98). In this adaptive co-management structure, teachers and students are equally empowered to adapt and evolve solutions, particularly as they relate to their learning needs and interests. Each contributes in their own way, from their own perspectives, often with insightful creative solutions to various challenges, so that collaboratively the diverse stakeholders in the educational community create conditions for thriving.

As noted above, in order to ensure all have an opportunity to work together to create optimal conditions for learning, there needs to be feedback between all stakeholders and levels in the system. As they are interdependently linked, information and perspectives need to cycle through and between all levels to enable a living learning system that can adapt and emerge as conditions change. As all thrive in sustainable living systems through adapting and emerging as conditions around us change, it is important to developing organizational systems and administrative approaches that are designed to support these transformative capabilities that lead to stability. This is what Senge et al. (2012) refer to as a *learning organization*; just what an educational system should be modelling.

We've seen how the structure of education based on ecological principles enables a panarchy organizational model, polycentricity, and adaptive co-management based on pluriversalism; recognizes the importance of collaboration, community, and diversity; and creates the conditions to enable a more resilient, dynamic system of learning. In order to bring this about, to support systems change, Dale (2002) recognizes the need to develop supportive policies.

The Whole Is Greater than the Sum of Its Parts

When the principle, "the whole is greater than the sum of its parts" guides organizational structure, administration, and leadership, the very ethos and structure of education support students, educators, administrators, and communities working together in helping create a sustainable, innovative, learning system. Once we develop these larger goals, this climate of education, and everything about our educational system, will transform to align with these goals (Meadows, 1999).

So rather than being focused on developing curricula, defining specific learning outcomes and standardized tests for all the separate subject disciplines for all grade levels, Ministries of Education need to focus on setting the conditions and structures to enable collaboration, transformational learning, complex adaptive systems (CAS) thinking, creativity, and innovation; enabling students, parents, teachers, and community members to respond as individual, community and societal conditions continue to change. As Ken Robinson (2016) recognized, we need to see education as an ecosystem:

> Schools have been obsessed with output, yield and have destroyed the culture of learning (the soil) and natural systems. We need to focus on soil, the natural process of learning we're trying to cultivate. Educational legislators need to understand their job is not command and control but climate control. We need to create the conditions conducive to thriving, learning, and growing. (49:11)

With the goal to mimic sustainable living systems based on an eco-centric paradigm, the traditional organizational structure transitions from a linear top-down hierarchy, focused on centralized control and siloed thinking, that maintains boundaries, to a dynamic, interacting panarchy.

Policy

Sterling (2016) is clear that we are faced with an unprecedented and huge learning challenge at every level, in which educational policy and practice need to play a pivotal role. The United Nations Economic Commission for Europe (UNECE) has been trying to advance education for sustainability with policy recommendations in professional development, and governing and managing institutions, as well as curriculum development and monitoring and assessment: "Within formal education systems these recommendations are addressed to policymakers, but they have implications for actors at all levels, including ministers of education, administrators, and educators. "All actors within education should take responsibility for the development of the Competences" (UNECE, 2012, p. 9).

In speaking to policymakers regarding the need to improve professional development for education for sustainability, the UNECE (2012) identify administration and leadership as key areas that need attention. They recognize we need to "Provide training and education in ESD for those in management and leadership positions in educational institutions. Leadership and management are key determinants of success in educational transformation at the institutional level" (p. 10). Once leadership and management recognize the importance and need to transform education for sustainability, the UNECE emphasizes a whole-institution approach should be adopted for the continuing professional development of educators in their workplace.

Senge (1990) has long advocated for schools to become learning organizations: "...organizations where people continually expand their capacity to create the results they truly desire, where new and expansive patterns of thinking are nurtured, where collective aspiration is set free, and where people are continually learning how to learn together" (p. 3). Through policies for and ongoing professional development, the competencies needed for sustainable educational transformation can be developed and supported. The culture and management of the entire organization are then supportive of sustainable development, and the whole educational system becomes a learning, adapting organization.

Beyond policies to support a whole-institution approach to professional development for Education for Sustainable Development (ESD), the UNECE (2012) recognizes policy changes are also needed to implement transformative management and leadership practices from the current linear, mechanistic model to a living system approach. This will support the development of necessary competencies they recognize are needed in developing sustainability.

Although outlined in the Chapter 1 earlier, these UNECE (2012) competencies bear repeating here as they relate specifically to various policies needed to transform management and leadership. The competencies are clustered around three essential characteristics of ESD: a holistic approach, envisioning change, and achieving transformation. The holistic approach includes three interrelated components: integrative, systems thinking; inclusivity; and dealing with complexities. Envisioning change covers competencies relating to the three dimensions of learning from the past; inspiring engagement in the present; and exploring alternative futures. Most importantly, achieving transformation covers competencies that operate at three levels: transformation of what it means to be an educator; transformation of pedagogy, i.e., transformative approaches to teaching and learning; and transformation of the education system as a whole (p. 16/17). UNECE (2012, p. 9) states, "These recommendations address not only Governments and regulators, but all decision makers and leaders who could have a role in providing frameworks, conditions and means for promoting the development of educator competences."

In developing supportive policies for governing and managing institutions based on the ecological principles of sustainable living systems, the UNECE (2012) recognizes the importance of developing and supporting a systemic, multi-scale panarchical approach that integrates regional to international educational initiatives:

Synergies among international, subregional, national and subnational processes should be identified and developed in order to facilitate the implementation of these recommendations. These will include synergies with existing processes, such as the Bologna process, Education for All and Life-long Learning, among others. Models of leadership that promote the enhancement of Competences should be developed and supported. Leadership is a key determinant of success in educational transformation at the institutional level; ESD requires the distribution of power across institutions in order to facilitate educational change. (p. 11)

Administration and leadership in educational systems need to create the conditions for these recommendations to take effect. Policies that create and support a culture of sharing of powers through adaptive co-management are essential. As the UNECE (2012) stated,

> Institutions and organizations that are involved in supporting learning should be encouraged to operate in ways and to maintain a culture that facilitates the development and practice of the Competences. This will include the way in which the organization distributes its own decision-making, manages its resources and conducts its relationship with the wider community. (p. 11)

Evaluation Policies

In developing ESD or sustainable education, what and how we evaluate can support or undermine transformation. Traditional centralized mechanistic evaluation, and the policies that support them, will undermine adaptive co-management processes based on a panarchical organizational structure. Instead, we need policies that support evaluating system processes as well as performance to ensure a resilient, dynamic, system. Focusing evaluation around the ecological principles to assess *interdependence, community, diversity, feedback loops, cycling* of ideas and information, *adaptation*, and *the ability to emerge* helps us know where to focus our evaluations. The UNECE (2012) recognizes,

> Audit and assessment as well as monitoring systems for educational institutions should be adapted or developed in order to assess the institution's contribution to sustainable development. Educational institutions should operate according to sustainable development principles as a contribution to ESD and create an enabling environment for the development and practice of the Competences.
>
> (UNECE, 2012, p. 12)

Adaptive co-management, as a key management approach in CAS (Armitage et al., 2009), involves working with multi-stakeholders across levels in

the educational panarchy and in the community; and policies and governance must ensure transparency and accountability. The UNECE (2012)
recognizes this will ensure legitimacy of ESD practices, as well as improve
and further develop the competencies.

In our traditional mechanistic systems we typically set targets and outcomes, for example, higher rankings on standardized league tables that
compare school in a region or a higher international ranking through
Program for International Student Assessment (PISA); or meeting budget
constraints to indicate a well performing system. However, once we shift
to CAS thinking what and how we evaluate also needs to change. The
UNECE (2012) recognizes, "Management should use evaluation as an
important learning tool that plays an integral part in strategic planning.
It should be seen as a reflection by all partners on ESD processes and
results" (p. 11).

To include all partners and use evaluation as a learning tool, complex
adaptive systems thinking process and learning goals become as or more
important than traditional product outcomes. To ensure an effectively functioning system that can learn, *adapt*, and *emerge*, that is, become resilient,
as conditions around various scales in the panarchy change, we need to
evaluate the strength and effectiveness of our relationships between actors
and scales in the panarchy (Plummer & Armitage, 2007). This is emphasized by the ecological principles of *cycling* and *feedback*. Administration
and leadership play a key role in developing feedback mechanisms and policies to ensure strong relationships are developed, pluriversality is honoured,
and there is effective communication to enable learning in adaptive co-
management, within and between scales in the panarchy.

Process and performance evaluations of both the system and stakeholders are key. We need effective policies that encourage people to explore
connections: the energy, dynamics, strengths and weaknesses of connections, as a valued part of the whole. When we replace the mechanistic root
metaphors of *top-down plan and impose* and *competition and survival of
the fittest*, by supporting *interdependence* and *community* with the positive,
non-threatening understanding of competition being a culture of striving
to be the best each of us can be through collaboration, we can then focus
on sharing and working together to try out innovations in a "safe-to-fail"
culture that welcomes feedback to encourage *adaptation* and *emergence*.
The challenge is getting people to feel safe in becoming life-long learners
and reflective practitioners. This will help administrators, teachers, students,
parents, and community members understand and develop awareness of
how to assess, think critically, and communicate what and how we are learning as a community of educators and learners. Supportive administration,
leadership, and policies need to support this process to ensure *the whole is
greater than the sum of its parts.*

In applying this living systems approach to administration and leadership, within each level in the panarchy, the ministry of education, the school district, the school board, administrators, and teachers will all adapt to align with the ecological principles so as to enable a more authentic, interactive, learning organization.

Ministry of Education

Being the largest scale in the educational panarchy, the Ministry of Education needs to liaise with international educational organizations, become an active participant in the various United Nations educational initiatives, exchange learning with international ministries of education, and work with other national level educational programmes and government ministries in developing a holistic understanding of societal needs and transformational opportunities regionally, nationally, and internationally.

At the local level, the Ministry of Education establishes and maintains the panarchical organizational structure, focused on society's goals to develop sustainable education where all can thrive. The Ministry needs to establish a learning organization culture of hope and inspiration based on the guiding ecological principles of *interdependence, community, diversity, feedback* and *cycling, adaptation* and *emergence* throughout all aspects of the system such that *the whole is greater than the sum of its parts.* As Robinson (2016) said, educational legislators need to establish the climate, create, and maintain the conditions conducive to thriving, learning, and growing.

To create this transformation, the Ministry of Education needs to set the policies that will enable systemic change and support a panarchical organizational structure based on CAS thinking and adaptive co-management. With this living system approach, the Ministry of Education's role is to stabilize, maintaining innovations that create the conditions to thrive; while receiving constant feedback from the other interdependent levels in the system and across society so as to embrace adaptation and emergence as conditions at various levels in the various systems respond and change (Holling et al., 2002). Dictating specific learning outcomes for each subject and grade level becomes unnecessary and counterproductive as learning will evolve and adapt based on local needs and contexts. This enables pluriversality, creativity, empowerment, and transformative teaching and learning at other levels in the panarchy that are context specific. The British Columbia Ministry of Education (2018) is moving in this direction in identifying core competencies and "big ideas" to guide learning and enable a personalized, flexible, and innovative approach at all levels of the education system, but it has yet to recognize sustainability and transition from its mechanistic organizational structure.

School District

A school district based on a Living Systems Framework is systemic, inclusive, democratic, participative, open, and responsive; has feedback loops between levels above and below in the panarchy; and builds and nurtures community. As an elected body of community representatives, working with appointed education professionals, the school district will also model an adaptive co-management approach based on pluriversalism and systems thinking so learning in schools is spatially, temporally, and contextually relevant; where decisions made at the district level are responsive and responsible to all other levels in the system above and below.

In recognizing the need for pluriversalism, it is important that school districts support *decentralization* so schools have the ability to self-organize and develop contextually relevant approaches to learning. Policies to support such decentralization can range from developing innovative outdoor experiential education programmes and alternative timetables to supporting a new approach to learning as exemplified by School District #68, Nanaimo Ladysmith, in a supporting Qwam Qwum Stuwixwulh School based on local First Nation ways of knowing and learning, through a *collaborative* partnership with the Snuneymuxw First Nation. To make pluriversalism work effectively *diversity* rather than "conformity" and "efficiency" are prioritized. As importantly, *feedback loops* between all areas and levels are established and maintained. School administrators, maintenance departments, teachers, community members, parents, and students from both elementary and secondary levels all contribute their opinions and perspectives in policy directions and decision-making.

The school district is a dynamic, innovative level in the system that puts the students' needs to learn in a caring, socially just, diverse, healthy environment as its priority. As such, it recognizes the child and community's interdependence with a healthy natural environment, thereby supporting learning in the community and natural environment by developing supportive policies and access to sustainable transportation. At the district level all purchases, procedures, and decisions are based on systems thinking, recognizing people as integral aspects of our natural environment, hence the importance of considering the socio-ecological impacts of all decisions for the health of the whole. Actions that do not support the health of the whole do not get approval – even if it is cheaper – as this will incur further social, environmental, or economic costs in the future, thereby modelling sustainability through its policies and decisions.

In providing for diversity, creativity, and innovation it's essential that the school district provides policy to ensure diversity of learning options and resources to maximize learning potential for all. Encouraging and modelling life-long learning and ongoing professional development at all levels in the network will enable further growth, *adaptation*, and *emergence* as everyone learns to thrive: adapting and emerging in response to changing conditions.

The School Level

Being an integral part of the community, policies need to enable schools and their resources to be administered as community schools that are available to all learners, of every age, so as to encourage life-long learning and professional development, in all aspects of education for sustainability. This aligns with UNESCO's Learning Cities initiative (UNESCO, 2017) that recognizes transitioning to a sustainable society needs to happen at all ages throughout communities, responding particularly to the targets of the Sustainable Development Goal 4, to "Ensure inclusive and equitable quality education and promote lifelong learning opportunities for all". In facilitating feedback loops between all levels in the nested system, policies are needed so that students and student representatives, as well as parents and the community, are involved in decision-making regarding school procedures, staffing and learning opportunities by voicing their concerns, opinions, and ideas. These policies will support transitioning from a dominator, top-down organizational structure based on centralized control, to an interdependent, interactive co-management model (see Figure 4.3).

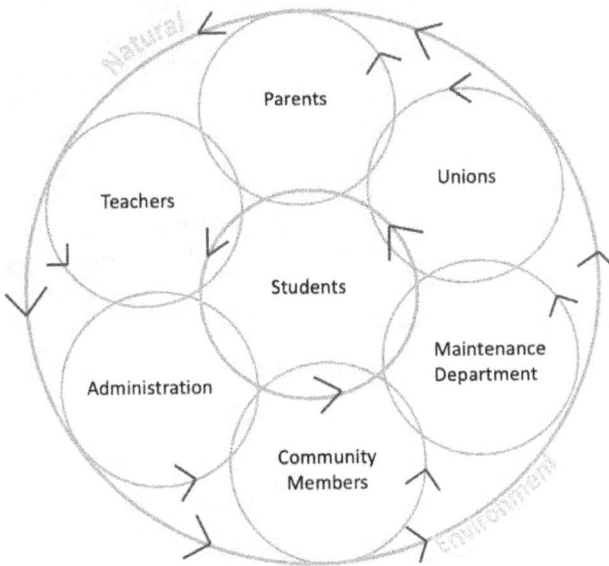

Figure 4.3 An Interactive Co-management Model. Created by the author.

Finally, school policies based on ecological principles can enable networking with other schools locally, regionally, nationally, and internationally to extend and support learning beyond the local community, developing extended learning communities to share ideas and innovations; and developing greater appreciation of our global living system built on *interdependence, community*, and *diversity*, recognizing *the whole is greater than the sum of its parts*, as we *adapt* and *emerge* in transitioning to a more sustainable society.

At the Classroom Level

Similarly, at the class level students, teachers and parents are involved in organizational systems that are open, responsive, inclusive, democratic, participative, learning frameworks. There is a very comfortable, relaxed, open atmosphere of partnership rather than domination, centralized control, and top-down imposition from teacher to students. The focus is on caring for adult/student relationships, honouring students' opinions and their learning needs as the number one priority.

Applying the ecological principles of *interdependence, community* and *diversity, adaptation* and *emergence*, in teaching and learning policies, students are empowered to be critical thinkers and be involved in making decisions regarding what they will learn, when they will learn, how they will learn, and in evaluating what they have learned. To enable diversity, adaptation, creativity, and innovation, the schedule for learning is open and may include special meetings with the school board, school administrators, or with teachers and fellow students regarding learning or the functioning of their class, what learning workshops teachers will be offering in a variety of classes, and what fieldtrips or outdoor activities may be organized. Students also need time to plan their day and be empowered to offer ideas on how they would like to learn. Throughout the organization of learning, students need to be honoured with respect, a belief, and trust in their abilities to learn and take responsibility as capable contributors to their learning, their community, and the natural world they are part of.

Systemic Effects

A Bioregional Approach

An example of applying a living systems management structure to education, from my PhD research, is an independent elementary school based on bioregional principles. Their vision is to provide educational opportunities that empower children to create fundamental social change towards more fulfilling and ecologically sustainable communities. Their bioregional philosophy and approach see humans as part of their bioregion, the natural world as a teacher and part of that community, recognize both the built and natural communities from the local to global scale, and provide freedom of expression and choice to students.

The organizational structure of the school was specifically designed to model and promote ecological principles of *community, interdependence, diversity, feedback,* and *adaptation* where participation, co-operation, collaboration, and positive synergies are sought to encourage empowerment of community members, administrators, teachers, students, parents, and volunteers.

The School Manual identifies The Bioregional Education Association's vision as a world in which:

- Communities are fulfilling, richly diverse and ecologically sustainable.
- All people are connected to and feel part of the natural environment.
- Children are connected to a diversity of people in their community.
- Children and adults act to maintain and create their communities.
- Diversity is tolerated, supported, and celebrated.

In working towards achieving their community-minded vision, the school's Mission Statement incorporates community statements such as, "we learn about our local bioregion, its natural & cultural elements"; "children, parents, teachers and mentors learn together"; "children and adults participate equally in age-appropriate decisions"; and "choices acknowledge responsibility toward the greater environment".

Speaking to the organizational structure of the school. The Principal said, "Our organizational structure is non-hierarchical so we seek an egalitarian model and we seek to have the children empowered as well in making decisions appropriate to their level." This was evident from the students' perspective as well. When the younger students were asked how ideas were developed at the school, one replied and they all agreed, "We have a meeting with the teachers, principal and volunteers and then if we think it's a good idea we do it." This egalitarian rather than top-down hierarchical management structure was very evident and consistently observed during all school visits. To specifically incorporate and solicit student, parent, and community opinions, Community Meetings are held from time to time. Students or any member of the school community can initiate these meetings so their concerns or ideas can be discussed. The students all felt they have had some input and say in these Community Meetings.

This organizational structure has been supported by consciously including all in consensus decision-making relating to the development of the school as well as the day to day running, curriculum development, and teaching/learning activities. A parent summarized this well:

> Decisions regarding curriculum and new initiatives are made co-operatively and by consensus. I saw it being lived and working when I came into the Board, that I felt was strong and successfully working co-operatively. I see it being lived in the classroom and in community meetings. Just the fact that there are community meetings shows that it works and is embraced throughout the school! It is empowering, I feel completely empowered here. I see parents come in here and volunteers and through their body language they seem empowered.

At a Board meeting, consensus-based decision-making was also a central focus. All persons present took turns speaking to the issues and had their views taken into consideration before a consensus was reached. A parent and Board Member confirmed this process by stating, "The teachers give their views from a teaching perspective, we talk about it as a Board, and the parents and students are asked their views on large paper where every-one can see other's views to build a consensus." This leads to collaboration. The Principal pointed out, "With consensus decision-making, your needs are met but not necessarily your wants."

This case study research showed that when diverse voices are heard a synergy, a strengthened community, and new levels of insight develop so schools and learning, appropriate to everyone's gifts and needs, can adapt and evolve as the whole becomes greater than the sum of its parts. Management and decision-making shifts from command and control, where solutions and innovations are planned and imposed from the top-down, to adaptive co-management approaches, based on systems thinking, where each level is responsive and responsible to all other levels in the panarchy. When educational administration is structured on this living systems model, *interdependence, diversity, community, feedback, adaptation*, and *emergence* happen naturally as foundational principles to enable and guide this transformation.

School District 64, Southern Gulf Islands, BC

The British Columbia (BC) Ministry of Education has developed a policy to empower school boards to develop strategic plans to improve educational outcomes for all students and improve equity for Indigenous students, children, and youth in care, and students with disabilities or diverse abilities. Collaborating with the Ministry of Education, School District 64 (Gulf Islands) identifies educational priorities that respond to local needs. These include a District Strategic Plan, First Nations, Métis, and Inuit Education Enhancement Agreement, Framework for Enhancing Student Learning,

Board Governance, District Programs, Financial Planning, Human Resources and Labour Relations, Learning Services, Facilities, Operations and Transportation, Information Technology, Communications, Anti-Racism Advocacy, and Climate Action and Environmental Sustainability.

Although School District #64 is situated within the traditional Industrial schooling system it is progressive in a number of ways in exemplifying some transformative approaches. To enhance student learning, the district's strategic goals are to inspire learning, integrate sustainability, and involve community. Collaborative district and school-based learning services teams work with parents and community partners to provide supports and services to meet the needs of all students. Numerous programs are offered by the district to strengthen diversity, community and interdependence.

The Indigenous Education Program is guided by an Indigenous team that includes Aboriginal Artist and Cultural Advisor, Indigenous Support Worker, Indigenous Youth and Child Care Worder, Outreach Team, Indigenous Ed Teacher Champions per site, and District Principal of Indigenous Education. SD64 partners with elders, knowledge keepers, youth mentors, and community for guidance and planning, as well as programming (SD 64, 2021).

Administration and leadership have also enabled interdisciplinary experiential learning within the traditional schools in the district as well as through specialist schools. In three elementary schools on the larger Salt Spring Island, parents and students have the option of learning in a multi-age class of K–3, based on outdoor experiential, emergent learning. As the district also includes a number of smaller island communities, they provide a diversity of sustainability educational opportunities to either learn on their island or travel to a nearby island via the water taxi.

Pender Hub (Middle Years and Graduation Program)

Pender Islands Elementary Secondary School provides Junior Secondary (Grade 8/9) programming for catchment students from Galiano, Pender, Mayne, and Saturna Islands. Salt Spring Island students wishing to explore a STEAM (Science, Technology, Engineering, Arts, Math) and eco-focused educational program are also welcome to attend. Being a smaller school (~140), students can experience a unique personalized learning experience and strong connection to staff and community.

Saturna Ecological Education Centre (SEEC)

SEEC is an award-winning school-of-choice where students live and learn together (boarding accommodations) on Saturna Island. SEEC embraces the development of youth through place-based and responsive experiential learning. In going outside the walls of the classroom,

SEEC ensures high school is a meaningful and authentic experience. For a full year of study, from Monday to Wednesday, students live and learn on Saturna Island. The program welcomes students in Grades 10, 11, and 12 to earn integrated course credits towards their individual graduation program. Projects combine science, social studies, physical education, and language arts to create a learning adventure that promotes critical thinking, social responsibility, and personal growth.

(SD 64, 2021, pp. 15–16)

Further enhancing diversity and community, the school district also offers programmes for adult learners, alternative programming based on individual needs, and a Connecting Generations Program to build bridges between schools and community by bringing together youth with adults of all ages, for conversations about a shared interest, skill, or life experience.

Finally, school gardens are an integral aspect of all school in the district, exemplifying *interdependence, feedback, adaptation,* and *emergence* through the programme:

Active veggie gardens thrive in every school in the Gulf Islands School District, where students learn, hands-on, about growing and eating healthy foods and caring for the earth. These gardens have been created by enthusiastic parent, community, and teacher/staff volunteers and are maintained by grant funds and donations.

These gardens connect families with our greater community, helping celebrate Island culture and transmit local agricultural skills across generations. Moreover, research has shown that teachers and children are happier learning outdoors at greener, tastier schools.

To support these gardens, a website was created at www.school garden.ca, featuring case studies and sharing resources and funding opportunities. Tools, seeds, and funds are also shared across gardens. Most years, school garden champions gather for "School Garden Learning Circles" to share successes and challenges, having already met at gardens on Mayne, Pender, Galiano, and Salt Spring over the past years. Garden-based Learning Workshops are also offered at District Pro-D days for Teachers. School gardens align with the District's goals of Sustainability and Community Connections, along with other strategic objectives.

(SD 64, 2021, p. 18)

Conclusion

As we've seen, using the ecological principles as a framework for the organizational structure, administration, and leadership enables the development of educational systems as dynamic learning organizations. Structured on the ecological principles of *interdependence, community, diversity, cycling, feedback loops, adaptation* and *emergence,* and an understanding of the *whole being greater than the sum of its parts,* organization and administration can get past the outdated command and control model to a framework where new significant polycentric developments in administration and leadership are encouraged and supported. *Interdependence* recognizes the vital roles all associated with schools can play at all scales, from students to parents, teachers, administrators, diverse community members, to the more-than-human environment. In creating a responsive educational system that can address individual learning needs as well as encourage collaboration and *adaptation* as conditions change, we develop a resilient, effective system of education that can respond to the rapidly changing needs of the 21st century. *Community* and *diversity* are key principles that further collaboration, responsibility, and resilience. *Cycles* of *feedback* create essential feedback mechanisms within and between levels in the educational system so it can constantly respond and *adapt* effectively. This dynamic learning framework is open and responsive to changing needs of both learners and society. It creates the conditions for students to thrive by encouraging the development of the whole person, encouraging systems thinking, creativity, innovation, life-long learning and *emergence.* In the following chapters we'll look more closely at how these same principles of sustainable living systems help support further transformations that are already underway in buildings, grounds, and resources; curriculum; and teaching and learning to enable the development of an education that will help us all thrive in the 21st century.

References

Armitage, D., Plummer, R., Berkes, F., Arthur, R. I., Charles, A. T., Davidson-Hunt, I. J., Diduck, A. P., Doubleday, N. C., Johnson, D. S., Marschke, M., McConney, P., Pinkerton, E. W., & Wollenberg, E. K. (2009). Adaptive co-management for social-ecological complexity. *Frontiers in Ecology and the Environment, 7*(2), 95–102.

Dale, A. (2002). At the edge: Sustainable development in the 21ˢᵗ century. UBC Press.

Grosfoguel, R. (2012). Decolonizing Western uni-versalisms: Decolonial pluri-versalism from Aimé Césaire to Zapatistas. *Transmodernity: Journal of Peripheral Cultural Production of the Luso-Hispanic World, 1*(3), 88–104.

Holling, C. S., & Gunderson, L. H. (2002). Chapter 2 resilience and adaptive cycles. In L. H. Gunderson & C. S. Holling (Eds.), *Panarchy: Understanding transformations in human and natural systems.* Island Press.

Holling, C. S., Gunderson, L. H., & Peterson, G. D. (2002). Chapter 3: Sustainability and panarchies. In L. H. Gunderson & C. S. Holling (Eds.), *Panarchy: Understanding transformations in human and natural systems.* Island Press.

Ireland, L. (2007). *Educating for the 21st century: Advancing an ecologically sustainable society* [PhD thesis]. University of Stirling, Scotland.

Meadows, D. (1999). Leverage points, places to intervene in a system. The Donella Meadows project: Academy for systems change. https://donellameadows.org/archives/leverage-points-places-to-intervene-in-a-system/

Nanaimo Ladysmith Public Schools. (2023). Qwam Qwum Stuwixwulh. Retrieve-from: https://qq.schools.sd68.bc.ca/

Olsson, P., Folke, C., & Berkes, F. (2004). Adaptive comanagement for building resilience in social–ecological systems. *Environmental Management, 34*(1), 75–90. https://doi.org/10.1007/s00267-003-0101-7

Pashby, K., da Costa, M., & Sund, L. (2020). Pluriversal possibilities and challenges for Global Education in Northern Europe. Retrieved from: https://files.eric.ed.gov/fulltext/EJ1281230.pdf

Plummer, R., & Armitage, D. (2007). A resilience-based framework for evaluating adaptive co-management: Linking ecology, economics, and society in a complex world. *Ecological Economics, 62*(1), 62–74. https://doi.org/10.1016/j.ecolecon.2006.09.025

Robinson, K. (2016). *Sir Ken Robinson & Dr. Peter Senge – education fit for the 21st century.* Disruptive Innovation Festival. Retrieved from https://www.youtube.com/watch?v=j1egRlszeH4

Gulf Islands School District 64 (SD 64) (2021). https://sd64.bc.ca/wp-content/uploads/2021/09/SD64-Operations-Framework.pdf

Schoon, M. S., Robards, M. D., Meek, C. L., & Galaz, V. (2015). Principle 7 – Promote polycentric governance systems. In R. Biggs, M. Schluter, & M. Schoon (Eds.), *Principles for building resilience: Sustaining ecosystem services in social-ecological systems* (pp. 226–250). Cambridge University Press.

Senge, P. (1990). *The fifth discipline: The art & practice of the learning organization.* Doubleday.

Senge, P., Cambron-McCabe, N., Lucas, T., Smith, B., Dutton, J., & Kleiner, A. (2012). *Schools that learn: A fifth discipline fieldbook for educators, parents, and everyone who cares about education.* Crown Business.

Sterling, S. (2016). A commentary on education and sustainable development goals. *Journal of Education for Sustainable Development, 10*(2), 208–213. https://doi.org/10.1177/0973408216661886

UNESCO Institute for Lifelong Learning (UIL) (2017). *Learning cities and the SDGs: A guide to action.* Retrieved from http://uil.unesco.org/lifelong-learning/learning-cities/learning-cities-and-sdgs-guide-action

UNESCO Institute for Lifelong Learning (UIL) (2023). *Webinar: Pluriversal literacies for sustainable and equitable education.* https://www.uil.unesco.org/en/articles/pluriversal-literacies-sustainable-and-equitable-education

United Nations Economic Commission for Europe (UNECE). (2012). *Learning for the future: Competencies in education for sustainable development.* Retrieved from https://unece.org/DAM/env/esd/ESD_Publications/Competences_Publication.pdf

Vlados, C. (2019). Change management and innovation in the "living organization": The Stra. Tech. Man approach. *Management Dynamics in the Knowledge Economy, 7*(2), 229–257. https://doi.org/10.25019/MDKE/7.2.06

Westley, F. (2002). Chapter 13 the devil in the dynamics. In L. H. Gunderson & C. S. Holling (Eds.), *Panarchy: Understanding transformations in human and natural systems.* Island Press.

5   The Hidden Curriculum
Buildings, Grounds, and Resources

What if you designed a school to work like a sustainable living system such as a bioregion, watershed, or tree, recognizing all aspects are components of interdependent communities, incorporating diversity in learning and feedback; where it generates all its own energy, cycles all its materials, adapts, and emerges new solutions as conditions change?

As we've seen, sustainable living systems have designed their support structures and systems based on the ecological principles of *interdependence, community, diversity, cycling* and *feedback, adaptation, emergence,* and *the whole is greater than the sum of its parts.* These principles are what help life thrive in a sustainable living system, enabling them to be resilient and adapt in response to their environments and communities. Designing a school based on these principles provides a comprehensive framework that pulls together and provides a context to support pedagogical and structural innovations; guides us in creating places of learning that support the needs we are recognizing in 21st-century learning; and creates an effective hidden curriculum that models and reinforces the importance of sustainability. So what would a school, designed like a bioregion, watershed, or tree look like in terms of buildings, grounds, and resources?

To begin this exploration, it's important to recognize buildings, grounds, and resources as the context, the environment if you will. Where learning happens significantly influences both the overt and hidden curriculum: it influences what type of learning is supported, and the values we promote in relation to our socio-ecological relationships. Our buildings, grounds, and resources reflect our educational philosophy, often unknowingly supporting the dominant status quo and subverting educational initiatives that are trying to foster an understanding of sustainability, as well as pedagogical innovations intended to foster more effective experiential, place-based learning (Ireland, 2007).

As we saw in Chapter 2, our typical schools and classrooms reinforce humans as separate from each other, our *community* and environment; that learning happens in decontextualized environments; and that

DOI: 10.4324/9781003389590-8

environmentally sound sources of energy and resources (such as renewable sources of energy, non-toxic furnishings, and cleaning products) are seen to be too expensive, and, therefore, not important or worth implementing. That is powerful teaching that undermines our efforts to transition to sustainable practices. How, then, do we create buildings, grounds, and resources so they align with and support pedagogical innovations that encourage the development of the whole child in relationship; creativity and innovation; and place-based learning that helps us develop more effective socio-ecological relationships and behaviours? In developing effective learning environments that align the hidden and overt curricula, we can use ecological principles to guide us.

Applying Ecological Principles

Interdependence

In designing indoor learning environments, the ecological principles of *interdependence* to ensure they are sustainably designed, creating healthy indoor learning environments, where the hidden curriculum supports the overt curriculum. Buildings need to model positive social-environmental connections to help create positive emotional connections with the natural world rather than separation from it. In developing deep, emotional connections with the natural world, buildings should be designed to be beautiful, giving one a sense of being part of the environment even when indoors, when there may be a need for shelter or warmth. This can help students develop a connection to and enjoyment of all seasons, regardless of the weather. Research has shown that environments that incorporate strong connections to natural environments powerfully contribute to our health and happiness, improving our cognitive, emotional, and physical well-being (Chawla & Escalante, 2007).

Community

Similar to natural systems, students are supported, adapt and learn best in communities they are a part of (Chawla & Escalante, 2007; Smith & Sobel, 2014). As such, schools need to dissolve their structural boundaries so that learning environments are integrally designed to be part of our human and non-human communities. When we model our interdependent relationships with our built and natural environments, as well as consider how we can encourage systems thinking, innovation, and creativity, it becomes clear learning environments need to support flexibility in learning and incorporate diverse place-based learning beyond the classroom. This is the network structure highlighted in Figure 3.2. In helping students connect with their communities, with their cultural and natural environments, we need to open our minds to designing learning spaces that are open, adaptable, and healthy so that students can learn effectively in the most appropriate place given the concepts, skills, and competencies they are exploring.

Incorporating the principle of community in how we design our buildings, grounds, and resources will help create learning communities. UNESCO developed the Learning Cities initiative to further enable our transition to a sustainable society by seeing cities as learning communities for learners of all ages (UNESCO Institute for Lifelong Learning, 2017). A focus on learning communities creates stronger, more extensive community-school connections, strengthening our understanding of interdependence, and the value of diversity. In doing so, schools will be available to all learners of every age encouraging life-long learning and community development. Students will be involved on a day-to-day basis in learning at a variety of community locations with community members, fostering intergenerational mentoring. Consciously designing schools to incorporate *interdependence, community,* and *diversity* helps ensure our redesign of education in the 21st century aligns with the principles of sustainable living systems; ones that create the conditions that allow students and communities to thrive.

Diversity

When learning happens in context, encouraging learning in community, this supports diverse, flexible place-based learning in the built community environment, in natural environments, as well as indoors. This diversity of learning options and resources maximizes learning potential for all ages. As such, learning happens in a variety of locations to situate learning by doing in context: in communities, parks, lakeshores, ponds, rivers, fields, forests, deserts, and/or mountains, as well as indoor learning hubs for the whole community. Such a learning hub could incorporate a library with IT facilities to connect people to larger learning networks, meeting rooms for multi-age groupings, classrooms, kitchens, art rooms, a woodwork shop, gymnasium, and theatre – anywhere learning is meaningful and contextually relevant. This development of a community school is already happening in many places where schools become flexible learning spaces and the heart of the community for all, exemplifying the core concept of interdependence.

Cycling

Sustainable building designs also incorporate the sustainable principle of cycling, where materials and nutrients cycle adding nutrients to the support system rather than waste and pollution. Applying this principle recognizes the need to capture natural lighting, passive solar energy, and rainwater, and to be ecologically designed in their construction and use of materials so they are non-toxic, as well as carbon-neutral, running on renewable energy. Furnishing and maintenance practices need to be non-toxic, locally sourced, designed to be reused or recycled, and designed to have a zero-carbon footprint. We are now learning how to mimic Nature to make regenerative, biodegradable materials using abundant natural building blocks. These materials effectively shift from waste and fossil based materials to regenerative ones that regenerate nature, support human health, and help develop a circular economy (Materiom, 2023; Stathatou, et al., 2023). Similarly, all cleaning products should be non-toxic and biodegradable. This teaches and recognizes we are an integral part of the environment, and our ecological footprints influence our health, happiness, and can positively contribute to the health of the whole.

Energy Flows from the Sun

As sustainable systems are powered with energy derived from energy that is currently being produced, similarly, our school buildings need to transition to renewable energy. This will allow them to become carbon-neutral and address government commitments to address climate change and cut running costs to be more economically viable. As importantly, we'll show students through this "hidden curriculum" that this is the way forward, and it is worth pursuing. In consciously transitioning to renewable energy, there are significant learning opportunities students should be involved in, connecting buildings, grounds, and communities with the overt curriculum.

Transportation is also an important consideration as this too is a learning environment imbedded in the "hidden curriculum." District transportation should be using renewable, non-polluting, carbon-neutral, socially just energy sources. Students who are not able to walk or cycle to school or on fieldtrips should travel on hydrogen or electric buses refuelled from renewable sources of energy. Electric buses are being used in a number of US states and Canadian provinces, creating economic, social, and environmental benefits which reinforce education for sustainable development and help synthesize the hidden with the overt curriculum as we transition to sustainability.

Feedback

In creating effective learning environments based on ecological principles, we need to stay open to constant feedback from all levels in the panarchy: students, teachers, administration, maintenance and support staff, parents, community members, and the natural environment. We need to look for and listen to the needs of all, becoming aware of unintended consequences, and new opportunities in looking for how buildings, grounds, and resources can better support sustainable education.

Adaptation and Emergence

To be resilient, effective learning spaces need to be able to respond to changing needs. Classrooms may need to be smaller or larger depending on the number of students and the space requirements of various learning activities. At times students may need indoor quiet spaces for reading and independent writing, larger spaces for collaborative learning that facilitates group explorations and discussions, and/or spaces that enable large gatherings, performances, and physical activities. Wherever possible, flexible learning spaces would have walls that can be shifted to accommodate various needs. Moveable glass walls and extended patio rooves can provide open access to the outdoors so natural light and the natural environment are part of the learning space with learning naturally flowing outdoors and in. In this way, buildings and grounds become integrated to support emergence of learning. Being responsive and flexible to accommodate change, and be open to facilitate outdoor learning, buildings can bring the community into the heart of learning spaces.

Ultimately, incorporating the ecological principles of *interdependence, community, diversity, cycling,* natural *energy flows, feedback,* and *adaptation* develops a context for learning that creates a more resilient learning *community,* enabling *emergence,* creativity, and innovation. It aligns the powerful hidden curriculum with the overt curriculum so as to support, rather than subvert, educational innovations in transitioning to sustainability, reinforcing we are integrally part of the environment and can develop effective socio-ecological relationships.

Practical Applications

A practising teacher and postgraduate student of Royal Roads University's Master of Arts in Environmental Education and Communication took on the task of designing buildings and grounds for a school district based on ecological principles. Andrew Delong's design work and conceptualization, below, exemplify how the hidden curriculum of buildings and grounds can be designed to support sustainable education.

Designing with Sustainability in Mind: A New Possibility

(by Andrew D.T. Delong, printed with permission)
    Instead of the current system for the construction of educational buildings, it is proposed to begin with Nature and build around it and with it (Figure 5.1).

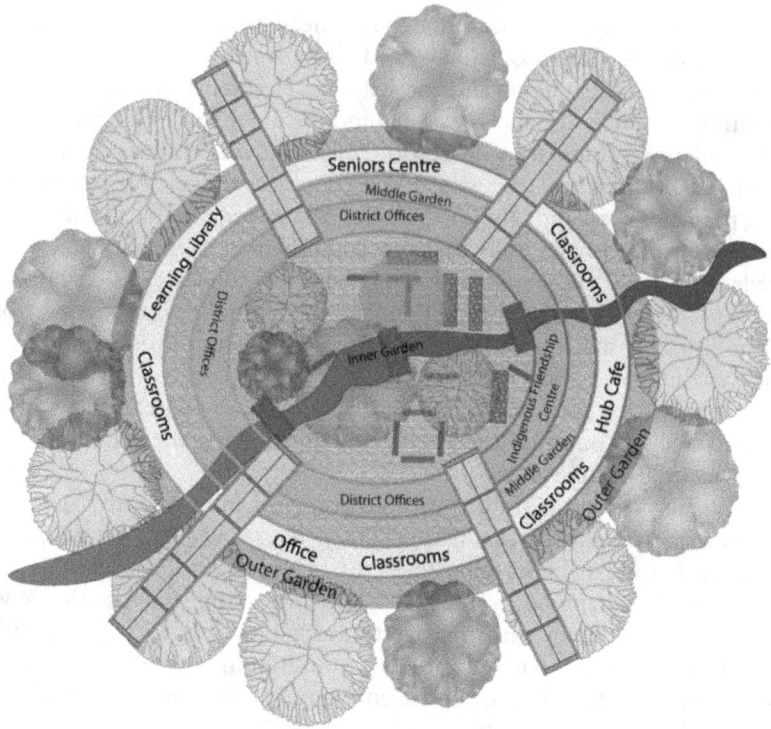

Figure 5.1 Designing with sustainability in mind: A new possibility.

Designed and reprinted with permission from Andrew D.T. Delong, Educator with School District 63, Master of Arts in Environmental Education and Communication candidate at Royal Roads University.

Based on the permaculture design principles, planning should begin with water (Mollison, 1988). Mimicking Nature, rainwater is "planted" so it can be harvested and used as many times as possible before returning clean to the environment as part of larger water cycles (Lancaster, 2019). In this way, rainwater is harvested and stored on site. Using the ecological principle of cycling, rainwater is filtered, used for potable needs, then filtered using plants, then used for toileting and kitchens, then filtered in biofilter systems, and finally used to irrigate trees and be fed back to the creek as clean safe water (Glendon BioFilters Canada Inc., 2022; Lancaster, 2019; Millison, 2021). The goal is that water is returned to the system cleaner than it entered after being reused multiple times along the way. The implementation of water harvesting and cycling of water demonstrates to students the interconnectedness of all living systems. Knowing that the school's water will eventually end up in the school stream encourages students to understand that there is "no away" (Lancaster, 2019).

The foundation of the building also rests on the ecological principle of energy flow. The sun's energy is captured through active and passive solar design and the use of wind turbines (Awasthi et al., 2018). The energy flow can also be seen through the photosynthesis of the trees – converting carbon into sugars – and releasing oxygen into the air allowing for a healthier environment for staff and students. It is based on a regenerative design model benefiting the natural world.

Building Design

The inner garden is a place built with water flowing through, ample vegetation, gardens with food to eat, and a wide variety of seating to enjoy the surroundings. This allows for opportunities to reflect, learn, and create community. This outdoor space allows for students and staff to have connections to the environment. It also allows the interconnectedness between the decision makers and the students.

Surrounding the garden are the district offices inspired by the Fuji Kindergarten (Tezuka Architects, 2007). Floor-to-ceiling movable glass windows allow decision makers ample opportunity to connect with Nature. Doors can be closed for private conversations or opened to allow natural airflow to occur. Windows are set back enough to have shelter from the elements but not feel disconnected from Nature. Departments work and collaborate instead of in individual silos. In larger districts, departments could be spread out to be connected to more schools. The balance would be found between connecting district staff with students and ensuring their work can be effectively done.

Consideration of where the facility is placed is also important. The main district facility would be placed in the lowest socio-economic area of the district. This would provide students with the least opportunities living in inner urban areas to experience Nature and the environment. Also, having what would be typically the highest needs and lowest funded school in the district together with the district office reinforces the importance of the decisions being made at the district level promoting the ecological principles of *community, diversity,* and *feedback.* For students, having access to natural outdoor space makes learning a transformative experience where they are immersed in the natural world and able to learn from the environment as a teacher.

Another garden surrounds the district offices and serves as another outdoor space for the classrooms and senior centres. The yellow section is an infusion of the community seniors' centre or senior's residence and the classrooms. This connection is built on the ecology principle of *interdependence* and *cycling* allowing for students to learn from elders and elders to learn from students providing a synergy of mentorships between students and elders. Students would have opportunities to participate in basic woodworking, weaving, gardening, storytelling, and reading with the seniors as their mentors. Students may work with elders to assist with using technology, setting up email accounts, and together they can fix broken electronics or small appliances. This allows for intergenerational information to be cycled in a meaningful way. Relationships with local Indigenous communities would be established to allow the opportunities for Indigenous storytelling and protocols to be used to help create a safe space for students and the community. To further allow this synergy to happen, working relationships between teachers and the seniors would occur to have the seniors drop into classes to learn and help.

The final set of gardens on the outside of the school serves as additional outdoor classrooms, meeting areas playgrounds, and the interface between the school and the community. This allows the school and district to provide a connection to Nature and the environment to people in the community who might have the least access. The outer gardens would extend the perennial herb and fruit tree gardens from the previous inner gardens. This food and medicine would be available for the community to access in the spirit of reciprocity for helping to look after the community. These gardens are a visual representation of living with the natural world. The school and district are better as a whole unit than separately.

Summary

In short, building schools that are built with Nature and not against Nature promotes schools to be more resilient and adaptative. Combining school district offices and schools while infusing opportunities to have the natural environment present allows schools to experience greater *community* and *diversity*. Connecting decision makers, community seniors and elders, and students in one building allows for interconnectedness and feedback from the people and systems involved. Together these elements of the educational system make for a much stronger whole unit than their parts would be separate. More research is needed to integrate these principles into retrofitting pre-existing schools to reach a similar outcome.

A Bioregional Approach

From my PhD research, the Independent Bioregional School's aim is to develop ecologically sustainable building infrastructure and grounds for learning and play, seeing them as part of the hidden and overt curriculum. Both the outside and inside environments have been thoughtfully considered and model ecological practices wherever possible. The playground has incorporated edible gardens, natural materials for building and playing with, composting and using rainwater, and a covered seating area so students can be outside in all weather. Expanding the gardens and building the soil from its original gravel cover has developed the school grounds. This was done through parents adopting a plot, student workshops in the spring, and help from staff and volunteers. Most of the gardening was done by a teacher although some enthusiastic students and parents helped with the planting and weeding. As a kitchen had not yet been organized at the school, a pumpkin and some potatoes were harvested in the autumn and the teacher and some parents took them home, made a pumpkin pie and potato dish, and brought them back for all to share at the Autumn Equinox Party.

As The Bioregional Education Association owns the building, the Board has total control over government regulations, and all felt they were empowered to influence the development of the learning environment. The older students felt they had input into changing the grounds and their ideas had been listened to. The younger students volunteered, "We are going to make a garden but not in the winter." Feedback from the parents on ideas for developing the grounds and use of building space was compiled at the Spring Equinox Party in 2003. Further plans to develop the outside area as a natural learning environment incorporating water, native plants, and habitats for animals; a water play area; a water stream where kids can circulate and pump back upstream by rowing; a bug study area; sitting area; playhouse; woodchip paths; a climbing tree; stepping stone paths; a granite boulder; fruit trees; a bench; and a decorated concrete pad for ball play, building toys, skipping, picnic tables, planters, a sand box as well as solar panels and windmills to get off the main electric grid. These plans show the active adaptations and ongoing development of the school. When asked if the grounds were ecologically managed, a teacher replied enthusiastically:

> Yes! No pesticides, fertilizers, and we add only soil that is said to be organic. The playground equipment is being replaced with outdoor gardens and natural climbing equipment. All materials are recycled and brought to school by bicycle. The edible gardens are watered from rainwater piped from collecting barrels.

The school founder recognized ecological management as one of the school objectives and built-in structures of the school. When asked about school supplies being made from sustainable resources, a teacher's response shows how conscious they are of sustainable resource use:

> We consider that for anything we purchase. We use almost exclusively paper that has been used on one side. We use tree-free paper for our photocopying. Mistakes in purchasing tend to get pointed out e.g. too many book purchases rather than using the local library. Any cleaners or soaps must be scent-free. We use peroxide bleach rather than chlorine bleach and use only natural chemicals in science experiments.

A parent recognized the efforts the school is making to control the potential negative impact of the hidden curriculum, and also making the students aware of their environmental impacts, by purposely not having garbage bins in the school. "The students must take their lunch garbage home, so they notice what garbage they generate; the environment and what impact they have is intrinsic in planning events."

Further sustainable energy and resource modifications to the building were planned as budgetary considerations could be met. Suggestions for the use of the building incorporated various learning stations such as a kitchen, an art/pottery area, a woodwork space, a library, a music room, and a science centre. There were also suggestions for specific workspaces for the lower elementary students and a separate one for the older elementary students. But even without these planned extensions, in good weather, the classroom doors were always open to the outside learning environment, and year-round, the students were free to move between the inside and outside environments depending on where they preferred to learn throughout the day.

This practical example shows what an intentional learning environment that models and teaches sustainability can look like. From the resources used in the building, for both maintenance and learning, to energy and how waste is dealt with, all become learning opportunities in the overt and hidden curriculum. The outdoor learning environments provide diverse opportunities for place-based experiential learning, even down to materials used, how they are sourced, and transported to the school. Beyond the immediate school grounds students are actively learning in the community whether that is taking advantage of the public library, learning in the biodiverse parks or in the community, or taking fieldtrips every Friday by bikes and trailer bikes so all can be mobile while keeping sustainability in focus.

Conclusion

In all educational systems, various levels in the panarchy have roles to play in ensuring the buildings, grounds, and resources support sustainable education. Working with the Ministry of Education, supported by finances and policies, school districts ensure all learning spaces and grounds are ecologically designed so that all materials used in the construction and furnishings are non-toxic, locally sourced; designed to be reused; designed for a zero-carbon footprint using non-polluting, renewable energy. Socio-ecological considerations also extend to all resources that support learning: teaching and learning resources as well as all cleaning and maintenance products are non-toxic and biodegradable; and all transportation uses renewable, non-polluting, carbon-neutral energy sources. Recognizing the importance of these interacting factors consciously develops a supportive hidden curriculum, reinforcing the importance of adapting to develop sustainable ways of living and interacting.

At this school level, there need to be a diversity of learning opportunities in a multitude of outdoor as well as indoor learning environments. Typically, we have designed schools to be separate from society, programming learning to happen Monday to Friday 9 am to 3 pm for students between the ages of 5–18. Yet learning is not really separate from play or daily living. Up until traditional school age, children learned all the time, no matter what they were doing, no matter where they were. Once "school" becomes a place where everyone in the community learns, associated with learning in context, at any time, the community and all its resources, both natural and human-made, become contexts for learning. With this interdependent model, it is not clear that the community and school are separate as children are involved in learning at a variety of community locations and community members of all ages are working side by side with children of mixed ages. This enables diverse learning resources and opportunities so learning can be developed through multiple intelligences, in a variety of contexts, creating multi-age classroom and/or learning opportunities.

These diverse examples recognize the importance of transforming buildings and developing learning spaces that integrate learning with the natural environment and community to foster more effective place-based, experiential learning. Having established how buildings, grounds, and resources can support sustainable education through the hidden curriculum, let's now turn to explore how the overt curriculum can be redesigned when based on the ecological principles of sustainable living systems.

References

Awasthi, A., Kumari, K., Panchal, H., & Sathyamurthy, R. (2018). Passive solar still: Recent advancements in design and related performance. *Environmental Technology Reviews*, 7(1), 235–261. https://doi.org/10.1080/21622515.2018.1499364

Chawla, L. & Escalante, M. (2007). *Student gains from place-based education.* Children, Youth and Environments Center for Research and Design, University of Colorado. Retrieved from http://www.colorado.edu/cye/sites/default/

files/attached-files/CYE_FactSheet2_Place-Based%20Education_December%20
2010_0.pdf

Glendon BioFilters Canada Inc. (2022). *The Glendon BioFilter system.* https://
www.glendonbiofilter.com/system

Ireland, L. (2007). *Educating for the 21st century: Advancing an ecologically sustainable society* [Doctoral dissertation, University of Stirling]. STORRE. http://hdl.
handle.net/1893/240

Lancaster, B. (2019). *Rainwater harvesting for drylands and beyond* (3rd ed.,
revised and expanded in color). Rainsource Press.

Materiom (2023). Vision for a Materials Revolution: Accelerating the transition
from fossil-based to regenerative materials. Retrieved from: https://materiom.
org/vision

Millison, A., Director. (2021, October 12). *How to recycle waste water using plants.*
https://www.youtube.com/watch?v=f-sRcVkZ9yg

Mollison, B. C. (1988). *Permaculture: a designer's manual.* Tyalgum, Australia,
Tagari Publications.

Smith, G. A., & Sobel, D. (2014). *Place- and community-based education in schools.*
Routledge.

Stathatou, P. M., Corbin, L., Meredith, J. C., & Garmulewicz, A. (2023). Biomate-
rials and regenerative agriculture: A methodological framework to enable circular
transitions. *Sustainability, 15,* 14306. https://doi.org/10.3390/su151914306

Tezuka Architects. (2007). *Fuji kindergarten | Educational buildings.* Fuji Kinder-
garten | Educational Buildings. http://www.tezuka-arch.com/english/works/
education/fujiyochien/

The Green School Bali. (n.d.). https://www.greenschool.org/bali/environment/

The Oberlin Project. (n.d.). https://oberlinproject.org/about/founder-visionary/
david-w-orr/#:~:text=In%201987%20he%20organized%20studies,change%20
on%20the%20banking%20industry.

UNESCO Institute for Lifelong Learning. (2017). *Learning cities and the SDGs:
A guide to action.* Retrieved from http://uil.unesco.org/lifelong-learning/
learning-cities/learning-cities-and-sdgs-guide-action

6 Curriculum

A Living Systems Framework

Some of the largest shifts in transitioning to sustainable education based on the ecological principles of sustainable living systems are in curriculum. In previous chapters we saw how the hidden curriculum can support or undermine sustainability education depending on the root metaphors it is built on. Our dominant educational system, based on the mechanistic root metaphors of *centralized, top-down control; efficiency; individualism;* and *separation of the economy, society, and the environment,* maintains the status quo through the hidden curriculum and are powerful contributors to the overt curriculum and how it is structured.

The traditional mechanistic system we have developed and cling to, being based on centralized control, and top-down planned and imposed learning outcomes, defines and structures curriculum in decontextualized, discipline-centered curriculum guidelines for each subject in separate grade levels. Assessment has been similarly structured to assess individual student achievement of these predetermined learning outcomes. We have developed a universal system where the government has taken ownership and control of learning to such an extent that it determines what to learn, when to learn, how to learn, and if a student has learned based on predetermined learning outcomes. It is all based on external control and moving students along a linear curricular conveyor from Kindergarten to Grade 12, and on up to teacher training.

We've learned that all species thrive when the conditions around them enable and support *adaptation* and *emergence* based on *interdependence, community, diversity, cycling,* and *feedback* – where *the whole is greater than the sum of its parts.* Humans are no different: curriculum in our educational systems needs to be open to enable and encourage *adaptations* and *emergence* – that is, learning – based on *interdependence, community, diversity, cycling,* and effective *feedback.* This enables curriculum to emerge naturally rather than by way of a decontextualized, linear, predetermined, and imposed process that typically lacks local or personal relevance at particular times in a student's development.

DOI: 10.4324/9781003389590-9

In rethinking curriculum at all levels, sustainable education needs to incorporate both intrinsic and instrumental sustainability-based curricula in developing a transformative educational paradigm commensurate with the social-economic-ecological challenges which face us (Sterling, 2010). In developing curricula that can support a sustainable educational paradigm and help all students thrive, we need a more organic, contextually relevant approach to curriculum that creates the conditions conducive to internally driven, transformative learning where students expand their innate talents and drives to learn, to adapt and emerge as they grow and change, and conditions around them change. In this way, curriculum needs to be systemic, contextual, interdisciplinary, and iterative, changing according to diverse needs in encouraging, and in response to, personal and societal transformations. This would mean students, teachers, and administrators, student teachers, teacher trainers, as well as Ministries of Education, would be involved in designing and enabling curriculum according to local interests, needs, and community contexts.

Curriculum Redesign from K-12 to Teacher Education

To ensure systemic change, this transformative educational paradigm needs to be the basis of teacher training curriculum as well as curriculum in K–12 education. When it is, the curriculum at post-secondary teacher training institutions allows teachers to develop sustainability as a state of mind, understanding from personal educational experiences what sustainable education is and how to facilitate it. Until systemic transformation of the educational system takes place, students coming into teacher education programmes will have been schooled into the industrial mechanistic paradigm, and expect to teach what they know and have experienced. Redesigning teacher training curricula based on ecological principles will enable transformative learning as teachers transition from the mechanistic educational system they were brought up in, to developing the competencies the United Nations Economic Commission for Europe (UNECE) (2012) identifies as foundational for educators so as to transition to sustainable education.

The various competencies identified by UNECE (2012) can help guide the topics for teacher training curricula. Teacher education needs to help teachers develop a holistic approach to education, envision change, and achieve transformation by incorporating competencies that help teachers learn to: understand sustainability; work with others to develop an eco-centric paradigm and new worldviews; facilitate transformative educational approaches; and embrace sustainability as a frame of mind and actions (UNECE, 2012). Each of the various educator competencies in each of these areas is addressed when we develop teacher education curriculum based on the ecological principles.

In addition to recognizing the importance of administrators and teachers developing competencies in ESD through professional development, as outlined in Chapter 4, the UNECE (2012) highlights the importance of using the competencies to develop effective curriculum documents teachers will work with in schools by noting,

> The Competences should be a basis for the review of curriculum documents...In order for educators to practice the Competences they should be supported by a curriculum which reflects such educational approaches. Textbooks and other educational materials should be reviewed to determine whether they reflect educational approaches suggested by the Competences. Materials may need to be developed to further support ESD.
>
> (p. 12)

Through this process, all aspects of curriculum need to be reconsidered: the hidden curriculum needs to foster intrinsic motivation, transformation, and developing sustainability as a way of thinking, while the instrumental, competency-based curriculum, from Kindergarten to teacher training, should lead to developing each student's ability to thrive and contribute effectively in the emergence of a sustainable society. To do this, let's look more closely at how the ecological principles, which give rise to sustainable living systems, can be used to frame how we redesign both the hidden and overt, the internal transformational, and instrumental curriculum, to enable sustainable education at all levels of the educational system so that students and society thrive.

Designing Curriculum Based on Ecological Principles

Interdependence

To organize and design curriculum to reflect how interconnected our natural, social, and economic systems are, rather than organizing learning around separate subjects, learning needs to evolve through integrated topics, relevant and of interest to students. Government Ministries of Education can give guidelines of topics they feel are important to develop competencies at various stages of development (rather than specific learning outcomes for each subject and grade level), how these are related, as well as the ecological principles that need to be understood through how curriculum is designed. Rather than defining specifics within separate subject disciplines, government guidelines would shift to broader systemic frameworks, big ideas, and "creating the conditions conducive to thriving, learning and growing" (Robinson in Robinson, 2016, 49:18).

For K–12 curriculum, the Alberta, British Columbia, Manitoba and Ontario Ministries of Education support these types of guidelines in their recently revised curriculum documents (British Columbia Government, 2018; Manitoba Ministry of Education, 2022; Ontario Ministry of Education, 2019). Although their curricula are still entrenched in mechanistic thinking by defining content in separate subject disciplines, it's encouraging to see the focus shifting to incorporate core concepts and cross-curricular competencies:

> British Columbia's redesigned curriculum brings together two features that most educators agree are essential for 21st-century learning: a concept-based approach to learning and a focus on the development of competencies, to foster deeper, more transferable learning. These approaches complement each other because of their common focus on active engagement of students. Deeper learning is better achieved through "doing" than through passive listening or reading. Similarly, both concept-based learning and the development of competencies engage students in authentic tasks that connect learning to the real world.
>
> (para 21)

Along with literacy and numeracy foundations, and concepts or "Big Ideas," core competencies are identified as sets of intellectual, personal, and social and emotional proficiencies that all students need to develop in order to engage in deep learning and life-long learning. The Alberta Ministry of Education (2011) emphasizes competency groupings, which contain descriptions of the attitudes, skills, and knowledge that contribute to students becoming engaged thinkers and ethical citizens with an entrepreneurial spirit. The three core competencies identified by the British Columbia Government (2018) are as follows: communication; creative and critical thinking; and positive personal and cultural identity, personal awareness & responsibility, and social responsibility. Alberta identifies core competencies of critical thinking, problem-solving and decision-making, creativity and innovation, social, cultural, global and environmental responsibility, communication, digital and technological fluency, life-long learning, personal management and well-being, and collaboration and leadership. Although these exemplify many of the competencies identified as necessary for 21st-century learning and by the UNECE (2012) for ESD, the ecological principle of *interdependence* brings them together, recognizing each is integral to thriving, yet it also highlights the need for further competencies in systems thinking, and the ability to adapt and change as conditions change. These interdependent core competencies along with key concepts can help Ministries of Education set the conditions for students to thrive recognizing they are interdependently connected to our diverse human and more-than-human communities through our socio-ecological relationships.

These curricular guidelines, being less prescriptive and subject-specific, enable more authentic interdisciplinary topic-based approaches. The British Columbia Government (2018) uses key concepts, principles, and generalizations to organize knowledge and solve problems within and across disciplines, allowing a more in-depth exploration of topics to gain deeper understanding and opportunities for the transfer of learning. They emphasize a concept-based curriculum allows for connections between big ideas rather than a list of topics to cover in isolation from one another. Howard (2020, p. 17) recognizes,

> 21st teaching and learning, like ESD, promotes progressive approaches. The radical change represented by 21st century education (Fullan & Langworthy, 2014) has led to project-based education and design learning. Authentic student-led design projects are recognized as powerful opportunities for creative, critical problem solving, and important skill development.

In a topic or concept-based curriculum, subject disciplines contribute diverse lenses to interdisciplinary explorations of big ideas and the development of competencies. Benchmarks related to each of the traditional disciplines are still important as they can inform developmental understanding, concepts, and competencies that are important in learning about and contributing to our world. But rather than form a disciplinary structure of education, they add interdisciplinary insights, developed through a diversity of topics.

So how do we translate this theory into practice? When I started The Green School, Scotland in 1996, I was faced with the question of what I was actually going to do when faced with students on Monday morning. To translate this model of education into practice, teachers need new ways to develop curricula and a framework to help bring learning to life. To do this I developed The Eco-Centric Curriculum Framework (Figure 6.1) that can be used and adapted to guide curriculum development based on local student and community interests and the ecological principles.

The Eco-centric Curriculum Design Framework, exemplified in Figure 6.1, shows how the ecological principles, as the core of how we educate and learn, become the organizational structure through which we can develop effective, meaningful learning opportunities. Topics, concepts, or "Big Ideas" form a holistic, central focus for curriculum development based on local, contextualized interests. These can be associated with science and social studies; while ways of expressing ourselves and our learning are drawn from language arts, physical education, music, drama, and art. As the ecological principles form the core of the curriculum, the emphasis

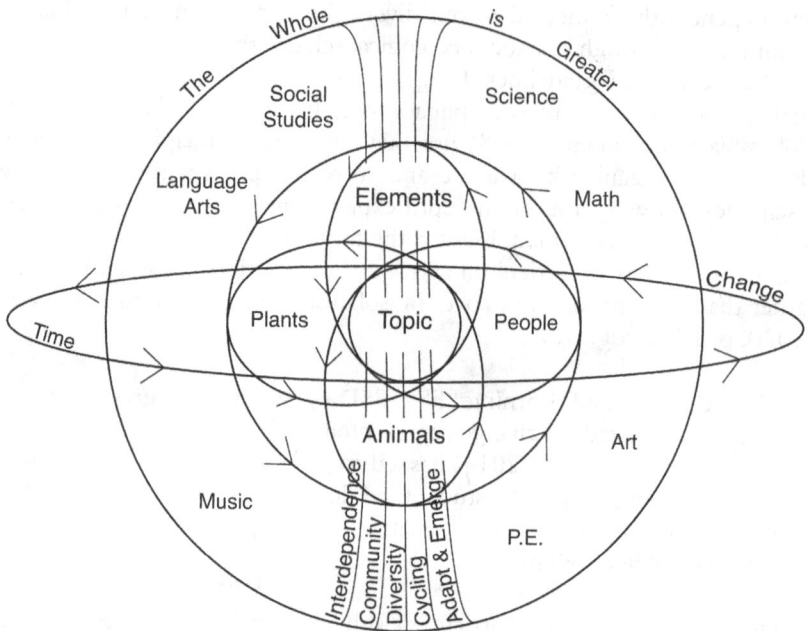

Figure 6.1 Eco-centric curriculum framework. Designed by the author.

is on developing systems thinking and seeing patterns that connect with the various subject disciplines adding diverse lenses and competencies to explore a particular topic or big idea. Learning and teaching through a project or topic-based, systemic curriculum, integrated within a community, highlights *interdependence, community,* and encourages and celebrates *diversity* to enable *adaptation* and *emergence* as our world changes over time. This leads to teaching and learning *the whole is greater than the sum of its parts.*

In developing an interdisciplinary topic-based unit of study, using the Eco-Centric Curriculum Design Framework, learning develops competencies, subject-based concepts, and understandings, as well as an innate understanding of the guiding principles of sustainable living systems. As the ecological principles of *interdependence, community, diversity, cycling, feedback, adaptation,* and *emergence* flow through the core of the hidden and overt curricula they further contextualize and extend learning, leading to a deeper understanding of dynamic, interdependent, complex adaptive systems we are a part of. This allows students and teachers to develop learning organically based on passions and topics of interest, and most importantly, it develops sustainability as a frame of mind (Bonnett, 2002) as the ecological principles are central to both the hidden and overt curriculum.

In developing curricula using The Eco-centric Curriculum Framework, holistic learning related to a topic of interest is developed collaboratively by students and teachers circling through the model, asking questions by looking at how elements (air, water, soil, nutrients, or abiotic elements), plants (including fungi), animals (including other terrestrial, air, and water species), and people are interrelated within that topic. Brainstorming with students brings up various knowledge and topics students and teachers are interested to explore further in each of these realms in developing a systemic understanding of the central topic.

Once this systemic mapping of a topic has been developed, weaving in various subject disciplines of science, social studies, maths, language arts, music, art, health & well-being, and physical education illuminates diverse, yet integrated, aspects of each topic enabling interdisciplinary understanding that *the whole is greater than the sum of its parts*, as well as holistic ways of expressing learning and developing competencies. This model can be used to co-develop curricula and make insightful operational decisions in how learning happens, considering how a particular topic or decision might influence our climate, air, water, plants, animals, and people, as decisions and learning are based on a holistic understanding of the world. In this way knowledge is not fragmented but is in reference to the larger context and its systemic characteristics. In the following chapter we will look at how teachers bring this holistic curriculum to life by focusing on critical thinking, ecological principles, experiential inquiry-based learning, reflective dialogue, and developing systemic thinking, becoming aware of how ideas and various disciplines are interdependent.

This curriculum design framework helps learners understand and explore the interdependent, dynamic Nature of our world as each topic unfolds. Considering how elements, plants, animals, and people interact for a particular topic helps students explore the diverse aspects of each topic while developing systems thinking. Layering in a consideration of the core ecological principles of *interdependence, community, diversity, cycling,* energy flows, *adaptation,* and *emergence,* as well as change over time in relation to a particular topic, helps students develop a broad, yet in-depth, systemic, contextualized understanding of the topic.

Community

With the ecological principle of community in mind, we consciously extend the curriculum to incorporate the local and extended communities, highlighting the larger communities we are part of. This systemic view means moving beyond individualism in how we design curricula. Through community and student involvement in developing curricular units, pluriversalism and diversity are emphasized and accommodated in how the curriculum can be adapted to local, diverse interests.

Students develop greater insights and make greater connections to reach deeper levels of understanding when curricular design encourages collaboration (Robinson, 2016; Sterling, 2001) and community-based learning. This focus on learning in community reinforces the community engagement highlighted in the previous chapter. By incorporating a sense of community, the curriculum needs to be designed to strengthen and give space for students to engage in community-based learning that can incorporate current issues and concerns, diverse cultures, perspectives, and ways of knowing (Ardoin, 2006). This place-based, experiential learning in local communities helps trainee teachers as well as K–12 students develop the UNECE (2012) competencies of exploring their personal worldview and cultural assumptions and seek to understand those of others and engage in real-world issues to enhance learning and make a difference in practice. By including community-based curricula in teacher training, teachers also learn to actively engage different groups across generations, cultures, perspectives, places, and disciplines, as well as including Indigenous knowledge and worldviews.

We know curriculum can also be extended through collaborations involving members of the community, parents or guardians, who may bring expertise and perspectives from their own lives and experiences. The British Columbia Government (2018) recognizes it is particularly helpful to cooperate and engage with experts from the community when learning about culture-specific contexts to avoid offence or misrepresentation or appropriation of culture. Cultural appropriation includes use of cultural motifs, themes, "voices," images, knowledge, stories, songs, drama, and so on without permission or without appropriate context or in a way that may misrepresent the real experience of the people from whose culture they are drawn, therefore collaborating with the community is essential. The British Columbia Government (2018) further notes that collaboration with community members exemplifies many of the First Peoples Principles of Learning (First Nations Educational Steering Committee [FNESC], n.d.) and nurtures cross-generational and relational learning. This curricular approach recognizes how interrelated and interdependent we are through the diversity of our communities in gaining insights through the various aspects of knowing, being, and becoming.

Diversity

As a foundation of curriculum design, we need to enable and enhance diversity in ideas, ways of knowing, ways to learn, and multiple intelligences: what we learn, when, where, and how. As sustainability and learning are processes rather than products, encouraging diversity of local initiatives and approaches would lead to enhanced critical and creative thinking, empowerment, and innovation in an ever-changing learning process that is responsive to changing local contexts. In this open, adaptive, curricular approach, diversity is not only honoured but also encouraged. Diversity within the curriculum is essential: the arts, hands-on activities, and critical and systems thinking that encourage imagination, innovative ideas, and empowerment need to be emphasized throughout teacher training and K–12. Diverse forms of knowledge (tacit, theoretical, technical, folk, encoded -in genes, language, cultural artifacts, plants, animals, poetic, spiritual, bodily) are all to be recognized within a systemic context and enabled through curriculum design. When we bring in pluriversalism, we also support diversity in types of schools and curricula as an eco-centric curricular approach embraces locally relevant adaptations such as French or other language immersion schools; First Nations curricula; special needs adaptations; and pilot schools experimenting with new curricular approaches.

Highlighting diversity in teacher training curricula also helps bring out ESD competencies identified by UNECE (2012). Exploring diverse worldviews and cultural ways of knowing and learning will facilitate the emergence of new worldviews that address sustainable development and encourage negotiation of alternative futures. Designing curriculum to recognize diversity helps learners clarify their own and others' worldviews through dialogue and recognize that alternative frameworks exist. To do this teacher training curricula needs to include how to work with different perspectives on dilemmas, issues, tensions, and conflicts.

Cycling

The ecological principle of cycling is inherent in this curricular approach at various levels. At the larger curricular design level, the fundamental concepts and big ideas are developed and reinforced through a spiral curricular approach that incorporates feedback from the community as societal needs change, and from all levels in the system. As students cycle through various topics at multiple age levels, they have an opportunity to explore new topics or revisit topics to allow for deeper understandings and seeing various concepts in new alternative ways.

At the most immediate level, highlighting cycling through curriculum design helps us develop circular, systems thinking rather than linear thinking. Highlighting how materials cycle through living systems to connect elements, plants, animals, and people in systems shows how *cycling* is central to the ecological principles of *interdependence, community, energy flows, adaptation,* and *emergence*. In this way, cycling also applies to how ideas cycle through an interdisciplinary exploration of various topics connecting diverse disciplinary lenses, building and expanding understanding in an iterative process of learning. This helps students understand the importance of this fundamental concept in learning to live sustainably in our world.

Feedback

In order to have a dynamic, organic curriculum that can respond to changing conditions, the ecological principle of feedback, both positive and negative, needs to be built into curriculum design through student-teacher co-creation of curricula. Students in K–12 and in teacher training also need to have open channels of communication built into the evolving curriculum to provide feedback and empower their involvement in how their learning unfolds. Through personal empowerment in curriculum design in teacher training, teachers will develop the competency evolving curricula with their students based on feedback and building on their own and their learners experience as a basis for transformation.

Adaptation and Emergence

As noted in the chapter on organization, administration, and leadership we have designed education around *efficiency, control, constancy, and predictability,* but in complex adaptive systems, that is, sustainable living systems, stability, and resilience arise from *persistence, adaptiveness, variability, and unpredictability.* In helping prepare students to engage effectively in our complex living systems, we need to enable the development of these characteristics and encourage students, from K–12 and teacher training, to embrace unpredictability so they can become comfortable adapting and emerging new understandings, competencies, and ways of being through how we design curriculum.

To do this, in this systemic, contextually relevant curricular model, local schools, teachers, students, and parents as well as trainee teachers in teacher training institutions need to be empowered to evolve, adapt, and change what and how they learn depending on how they assemble learning opportunities and changing circumstances. If education becomes locally relevant it will mimic innovation in sustainable living systems, enabling innovations and responses to local conditions to self-assemble or emerge from the ground up, devolving prescriptive top-down, centralized control in favour of local needs and priorities. Contextually relevant, deep learning will lead to a more decentralized, student and community generation of topics. By allowing locally related adaptations related to the big ideas or general topics, education becomes more stable and resilient as schools embrace *persistence, adaptiveness, variability, and unpredictability.*

We need to be able to expect and embrace the unexpected, seeing unpredictable occurrences as true and positive learning opportunities. Engaging with and embracing variation, change, adaptiveness, and persistence is to embrace learning. These qualities that lead to life-long learning are not overtly taught but are developed through a curriculum that is open rather than predetermined with pre-defined learning outcomes and lock-step units and lessons in each subject discipline. Authentic learning and empowerment come from developing these capabilities in becoming a learner that can explore the unknown, responding effectively with inner motivations and capabilities to adapt as an innate learner. This type of responsive curriculum is both internal or transformational and external or instrumental as it is open, based on developing competencies that embrace variation, change, adaptiveness, and persistence through inquiry-based learning.

These are the qualities and competencies students as well as teachers need in our rapidly changing world so teacher training curriculum needs to be similarly designed. *Adaptation* and *emergence* support the UNECE (2012) competencies for educators being able to critically assess processes of change in society and envision sustainable futures; facilitate the evaluation of potential consequences of different decisions and actions; and communicate a sense of urgency for change and inspire hope. Becoming comfortable with *adaptation* and *emergence* through their own learning will help students at all levels develop internal motivation and develop further competencies to make a positive contribution to other people and their social and natural environment, locally and globally. Through teaching and learning through adaptation and emergence each life-long learner be willing to take considered action even in situations of uncertainty, and challenge assumptions underlying unsustainable practice (UNECE, 2012). Curriculum designed to enable *adaptation* and *emergence* in teacher training and in turn from K–12 leads to transformative a*daptation* and *emergence*, understanding *the whole is greater than the sum of its parts*.

The Whole Is Greater Than the Sum of Its Parts

There is something special, something magical that happens, when students see the gestalt, realizing the bigger picture, seeing how various parts of the whole are relevant and interrelated, giving rise to a deeper understanding and meaning. We often lose these larger deeper meanings when we get so immersed in breaking curriculum into bits and pieces, organized in a linear, decontextualized format to try to make teaching and learning efficient and predictable. By teaching in a linear style, moving from objective to objective and consequently thinking in a singular direction, it's harder to draw bigger connections and focus is drawn to a reductionist examination of parts. Students rarely see the larger meanings and significance, becoming lost in a series of lessons focused on various parts and pieces; reflected in one of the most common questions and statements teachers hear students asking, "Why do we need to learn this? It doesn't make sense."

This simplistic mechanistic thinking in disciplinary silos maintains the status quo compromising both critical and systems thinking needed to thrive with the challenges we face in developing a sustainable society. It

creates a linear, mechanistic way of thinking about the world as disconnected parts, with the false notion that society, the economy, and the environment are separate. This mechanistic thinking that science, maths, social
studies, language arts, and the fine and performing arts are separate is further reinforced by not only developing separate curriculum units for each
subject discipline but also having them each taught separately, by separate
teachers, and assessed independent of each other. This is further exagger-
ated in post-secondary education where the majority of students and educators focus almost exclusively in either the arts or sciences; in social sciences
or natural sciences, not seeing how these various subjects are disciplinary
lenses contribute to our seeing and understanding the whole. Senge et al.,
in *The Necessary Revolution*, state:

> A sustainable world will only be possible by thinking differently. With
> nature and not machines as their inspiration, today's innovators are
> showing how to create a different future by learning how to see the
> larger systems of which they are a part.

(2008, p. 10)

When looking at how things in our world works, from Nature to human interactions, issues and possible solutions, we find a constellation of interacting
parts and a series of complex actions and consequences affecting a greater
whole. Being able to do this needs to be a key competency of educators
(UNECE, 2012). Learning is far more than teachers or K–12 students being presented with various separate disciplines, broken down into subjects,
units, and lessons carefully sequenced using either a cognitive process, focused on our left brains in science, maths, or language arts; or our creative
right brains in a few arts classes; or by taking time out to exercise our bodies
in PE classes. Curriculum needs to address the whole person in synergy,
expanding its dominant emphasis on efficient transmission of cognitive skills
and subject-centred knowledge in decontextualized bits and pieces. It needs
to incorporate knowledge that is understood as multidimensional and interconnected within a larger context; to explore a diversity of attitudes, values,
intuition, and multiple ways of knowing; and encourage a sense of empowerment and active engagement of the whole person in becoming and contributing to the larger socioecological systems of which we are an integral part.

Our world in the 21st century has changed and continues to change
at significant rates, with digital communications, artificial intelligence, the
rise of circular economies, decolonization, pandemics, biodiversity loss,
and climate change. We have found our outdated, mechanistic curricular
approaches must also change if we are to be able to respond and think
systemically as well as critically; to synthesize with new insights through
collaboration, creativity, and innovation.

Focusing on the ecological principle of *the whole is greater than the sum of its parts*, curricula for teacher training, as well as K–12 levels, need to provide space to consider the bigger picture through systems thinking. Particularly for teacher training but also in increasingly higher levels in the K–12 system, the curriculum needs to include an examination of the root causes of unsustainable development and the urgent need for change from unsustainable practices towards advancing quality of life, equity, solidarity, and environmental sustainability. In seeing sustainability as an evolving concept, teachers and students need numerous curricular opportunities throughout their education to connect sustainable futures and the way we think, live, and work (UNECE, 2012). This type of curriculum is inquiry based, not prescriptive, so as to enable their own thinking and action in relation to sustainable development. This opens opportunities for transformation through systems thinking and explorations of personal, cultural, and societal norms. Recognizing *the whole is greater than the sum of its parts* helps us develop curricula based on holism. Ultimately, this leads us to considering and understanding these larger systems we are part of, and the communities in which we interact.

As teachers are a product of the traditional mechanistic educational system, teacher training curricula also need to consider the larger educational system we are part of and its mechanistic roots that help maintain the status quo. This is essential so they come to understand why there is a need to transform the education system, the way we teach and learn, and why it is important to prepare learners to meet new challenges. Having curricular opportunities to explore the larger panarchical system they are part of will help teachers think in systems recognizing how they are influenced by other scales in the educational system and they, in turn, can be part of adaptive transformations. This will help educators challenge unsustainable practices in their own institution and across educational systems.

Putting the Principles into Practice

While using the ecological principles to support a transformative process of learning through the hidden curriculum, the overt curriculum, developed through the Eco-Centric Curriculum Design Framework (Figure 6.1), puts all the principles into practice in developing a holistic, contextual understanding. Take the topic of energy for example. By exploring energy through the main components of natural living systems, students will develop a systemic understanding of this important topic in transitioning to a sustainable society. This topic that can be explored through a variety of avenues, at many different levels depending on community contexts and the levels of development and understanding of the students.

Cycling through the Eco-Centric Curriculum Design Framework, an energy curriculum emerges by exploring the various aspects of an ecosystem while bringing out the ecological principles throughout. Looking at energy in terms of the basic elements, water, air, fire, and the sun, provides extensive explorations from understanding what energy is to renewable forms of energy, how we can store energy, and climatic changes based on the use of various energy sources. Connecting and extending curricula to energy in relation to plants and then animals, students and teachers can explore how these aspects of our ecosystem access and use energy, contribute to energy cycles, and how their energy systems are interdependent, diverse, and adapt to changing conditions. In turning to people, students can then explore where we source energy, how we use it, a variety of issues we are facing, and how we can learn from Nature to adapt and emerge new sustainable solutions.

Considering and bringing in the ecological principles of *interdependence, community, diversity, cycling, energy flows, adaptation,* and *emergence* connects all four quadrants, helping students see how all these ecological principles and various aspects of the ecosystem contributes to the Earth energy systems. Integrating the various disciplinary lenses extends and deepens an understanding of energy as an integral aspect of our socioecological relationships, and how addressing environmental, social, and economic imperatives is important in developing sustainability as a frame of mind. By weaving in ecological principles and diverse perspectives of social studies and science, maths to explore relationships, and language arts, PE, and the fine and performing arts to access and express learning, the Eco-Centric Curriculum Design Framework helps teachers and students develop various levels of inquiry, according to developmental needs and interests. Yet it also develops systems thinking and an understanding of our world as an integrated complex adaptive system with multiple perspectives and influences.

Practical Application: The Green School, Scotland

The curriculum for The Green School in Scotland evolved using this eco-centric curriculum framework through collaboration with the students and myself, their teacher. To start the learning in autumn, the topic of food was agreed upon as both timely and of great interest to all. Cycling through the framework, the teacher facilitated a brainstorming session so students could share all their ideas and interests in food related to elements, plants, animals, and people. Numerous *who, what, when, where, why, and how* questions came up and were recorded to help inspire inquiry-based learning opportunities. As students from ages 5 to 12 attended the school in mixed-age groups, the topic-based, eco-centric approach enabled the incorporation of diverse interests, levels of learning and development. This initial brainstorming session created a holistic plan for learning in all these areas so the next step was to develop experiential inquiry learning activities that would highlight the various ecological principles through integrating the various disciplinary lenses for learning at various levels.

Starting with elements developed curricula to look at where food comes from, what is needed for plants to produce food, and what they need to grow: nutrients in soil, air, sunlight, and photosynthesis. As it was autumn, inquiries included changes in weather, temperature, and available sunlight. Ecological principles helped structure learning and were highlighted throughout: *interdependence* between elements as well as between elements and plants; *diversity* of elements in various locations; *cycles* in seasonal changes; *feedback* plants get from changes in the elements; *adaptations* due to climate changes (such as drier soil due to drought), plants' *adaptations* seasonally; and *emergence* in producing seeds, nuts, and fruit. These curricular explorations integrated elements and plants, and at various levels of depth and detail for various students, science, geography, maths, language arts, as well as creative dramatizations and physical education.

The curriculum then moved into animals, looking at where animals get their food, whether they are herbivores, omnivores, or carnivores; *interdependence* between elements, plants, and animals; how they make up a *community* ecosystem; *diversity* of animals based on food sources; and how seasonal *cycles* lead to food adaptations. Again, competencies and further understandings were developed through the lenses of science, maths, geography, language arts, music, and art.

Focusing on people brought out opportunities to explore food in a multitude of ways: how interdependent we are with elements, plants, and animals for our food; foods we grow, gather, and eat in our families and communities; the importance of diversity types of food in our diet; how we grow, cook, and preserve different cultural foods; our food preferences based on tastes, smell, and feel as well as ethics; seasonal foods and eating locally; and how we adapt and emerge variations in our food systems based on climate changes and sustainability. With explorations focused on food and people social studies comes to the forefront as well as science, health, maths, language arts, music, art, dance, and physical education.

As there are so many avenues to explore, the teacher offers teaching and learning activities to bring curriculum competencies to life. As there are so many levels that can be explored throughout the unit, the teacher developed the curriculum and facilitated learning according to student interests and needs, intentionally weaving in various subject disciplines to develop competencies and expand learning for various levels. For example, when investigating plants and elements, using a scientific lens younger students would be doing experiments to see how much sunlight and water a plant needs, while older students were learning about photosynthesis and creating models to demonstrate the chemical process. With mathematics some younger students were sorting seeds, using addition and subtraction, while older students were graphing different seed ratios of various plants. As the learning is inquiry-based, all explorations provided opportunities to develop curriculum competencies in ecological literacy; communication and media literacy; information technology; critical thinking, problem-solving, systems thinking, innovation, and creativity; authentic real-world learning; health and well-being, adaptability, and life-long learning.

Following students' interests in learning, the curriculum developed organically, with one topic leading into another. Within our topic of food students often elected to follow up on various subtopics of particular interest, individually or in smaller groups. Some chose to look at permaculture, or a healthy vegan diet, while others were fascinated with what dinosaurs ate. All students took turns helping design and prepare healthy lunch meals with the help of a grandmother so we could eat together in a community. With some individual projects still happening, collectively we eventually shifted into exploring the topic of Winter as the season changed, and then Spring with various specific topics of interest becoming the focus of the curriculum development. The students even initiated a school garden, collecting compost from the school kitchen so as to create the right soil conditions, strategically placing it on the grounds to take advantage of sun and easy of watering, and designing how it could be adapted to keep the local rabbits away. In this way, the learning continued and the curriculum developed organically based on contextual relevance and student interests. Holistic topic-based learning led to the students developing deeper, systemic understandings of ecological principles, 21st-century competencies, and subject disciplinary knowledge and skills while exploring topics that interest them. In learning holistically about complex, dynamic, adaptive food systems, students naturally saw how various disciplinary lenses and aspects of food systems lead to the whole being greater than the sum of its parts.

Conclusion

As we've seen, the ecological principles create the foundation for developing a curriculum conducive to students strengthening their innate desires to learn, their creativity, and abilities to engage in the world to their full potentials by developing the conditions for students to thrive. This design framework, being open-ended, encourages teachers and students to be involved in curriculum development that encourages creativity, diversity, and innovative teaching and learning that can be responsive to local contexts, needs, and interests. In this way it develops extrinsic, instrumental, as well as intrinsic, transformative competencies that enable sustainable education (Sterling, 2010). Inquiry learning, innovation, and creativity rise to the forefront as students explore diverse topics and their interrelationships. Being interdisciplinary it creates a curriculum that supports rather than subverts systems thinking. It breaks free of the counterproductive status quo of the Industrial Era's mechanistic hidden and overt curriculum, based on linear thinking and top-down control. In doing so, it aligns the hidden curriculum with the overt curriculum, and with the systems thinking needed in the 21st century to help develop sustainability as a frame of mind, as a way of seeing and interacting in the world.

References

Alberta Ministry of Education. (2011). *Framework for student learning: Competencies for engaged thinkers, ethical citizens with an entrepreneurial spirit.* Retrieved from https://open.alberta.ca/publications/9780778596479

Ardoin, N. M. (2006). Toward an interdisciplinary understanding of place: Lessons for environmental education. *Canadian Journal of Environmental Education, 11*(1), 112–126.

Bonnett, M. (2002). Education for sustainability as a frame of mind. *Environmental Education Research, 8*(1), 9–20.

British Columbia Government. (2018). *BC's new curriculum: Curriculum overview.* https://curriculum.gov.bc.ca/curriculum/overview

First Nations Educational Steering Committee (FNESC). (n.d.). *First peoples principles of learning.* Retrieved from https://www.fnesc.ca/wp/wp-content/uploads/2020/09/FNESC-Learning-First-Peoples-poster-11x17-hi-res-v2.pdf

Howard, P. (2020). Living schools and 21st century education: Connecting what and how with why. In C. O'Brien & P. Howard (Eds.), *Living schools: Transforming education.* Education for Sustainable Well-Being Press (ESWB Press).

Manitoba Ministry of Education. (2022). *Curriculum kindergarten to grade 12.* Retrieved from https://www.edu.gov.mb.ca/k12/cur/science/index.html

Ontario Ministry of Education. (2019). *Education that works for you.* Retrieved from https://news.ontario.ca/en/backgrounder/51527/education-that-works-for-you-modernizing-learning

Robinson, K. (2016). *Sir Ken Robinson & Dr. Peter Senge – Education fit for the 21st century.* Disruptive Innovation Festival. Retrieved from https://www.youtube.com/watch?v=j1egR1szeH4

Senge, P. M., Smith, B., Kruschwitz, N., Laur, J., & Schley, S. (2008). *The necessary revolution: How organizations and individuals are working together to create a sustainable world.* Doubleday.

Sterling, S. (2001). *Sustainable education: Re-visioning learning and change.* Green Books.

Sterling, S. (2010). Learning for resilience, or the resilient learner? Towards a necessary reconciliation in a paradigm of sustainable education. *Environmental Education Research, 16*(5–6), 511–528. https://doi.org/10.1080/13504622.2010.505427

United Nations Economic Commission for Europe (UNECE). (2012). *Learning for the Future: Competencies in education for sustainable development.* Retrieved from https://unece.org/DAM/env/esd/ESD_Publications/Competences_Publication.pdf

7 Teaching and Learning
Empowering Change

Having looked at how organizational structure, administration and leadership, buildings, grounds and resources, and curriculum can provide the support structures for sustainable education, we can now look at how teaching and learning based on ecological principles bring sustainable education to life. Teaching and learning that are guided by the ecological principles of *interdependence, community, diversity, cycling, feedback, adaptation, emergence,* and *the whole is greater than the sum of its parts,* develop systems thinking in teachers and students, and the ability and motivations to be fully present and active in transformative, sustainable education. They also support the United Nations Economic Commission for Europe's (UNECE's) (2012) identification of educators needing to: be both a facilitator and participant in the learning process; a critically reflective practitioner; inspire creativity and innovation; and connect the learner to their local and global spheres of influence. This is contextually relevant teaching and learning that brings out empowerment, innovation, and creativity. As we will see below, each of these principles leads to and support the progressive pedagogy that is at the heart of teaching and learning to enable inner transformation and the development of 21st-century competencies.

In considering teaching and learning, it is important for educators as well as teacher trainees to keep in mind the mechanistic roots of our present educational context teachers have been educated and trained in, work in, as the influential mechanistic paradigm we unconsciously carry with. These can influence our approaches to teaching and learning, as well as our lack of openness to innovative sustainable education approaches – without our realizing it. While you are reading through this chapter, try to be aware of where these limiting mechanistic root metaphors arise for you. Try to recognize where you are breaking these ideas into smaller parts, and where you are connecting these ideas into larger wholes. In engaging with these ideas, we often fall back on feelings that we need centralized control, or summative testing and grading to sort students – in assuming the mechanistic root metaphor "competition and survival of the fittest" will motivate

DOI: 10.4324/9781003389590-10

students, and lead to effective teaching and learning that ensures students can progress along a linear educational trajectory "to become successful". In transitioning how we teach and learn, so as to support the transition to a sustainable society, we need to recognize when and where these mechanistic root metaphors are at play – and question these assumptions. This is an important part of the process of unlearning reductionism, letting go of mechanistic root metaphors, freeing ourselves of the restrictive cocoon we are typically unaware of, so as to open ourselves to teaching and learning that fosters resilience in transitioning to a sustainable future.

As previously discussed in Part I, in our traditional educational system, a teacher is trained to focus on prescribed learning outcomes and competency indicators for each subject at every grade level. Although teachers are expected to take students' individual needs into account, the system is not designed to cater to their interests or collaborate with them beyond a very superficial level. This tends to add significant stress to teachers who are trying to cope with diverse levels of abilities and interests. It is extremely challenging to be inclusive and provide meaningful learning activities for all when students can range from those that might have cognitive, emotional, physical, and/or linguistic challenges to those who are gifted in a multitude of ways.

While trying to address all these challenges in the mechanistic system they are immersed in, teachers reinforce the mechanistic mindset and status quo through the typical daily teaching and learning routine that is mechanized and preprogrammed through top-down control, based on separate discipline subjects. Even at elementary grades where the classroom teacher has control over what is taught, when, and how, they replicate a mechanistic timetable (Ireland, 2007). When the bell goes, it is time to stop thinking about science, for example, and start an unrelated social studies or language assignment – or be evaluated – whether students are ready or not. What is left for the student to decide? Students who try to make their own decisions as to what, when, or how they want to learn become problems as they do not fit easily into the system with their diverse needs. As a result, they are either be forced out, voluntarily leave, or are brought back in line, forced to conform, or have their needs partially met. The student has little to no say about what, when, or where they learn or even if they have learned.

The industrial foundation of education, illuminated by root metaphors of the dominant paradigm (*the world works like a machine and individuals are independent units of that machine; society is best controlled through centralized, top-down control; the economy, society, and environment are separate; the world is in our hands – humans can and have the right to control the environment; nature is a resource; rational thought (neutral, natural, and culture free) is the epitome of intellectual achievement; take the world apart to understand it; progress is linear with knowledge comprised of separate linear building blocks;* and *survival of the fittest*), clearly frustrates efforts to develop systems thinking, innovation, and creativity, as traditional pedagogical approaches based on these mechanistic root metaphors subvert rather than support nonlinear transformative learning (Ireland, 2007; Robinson, 2015; Sterling, 2001).

As discussed earlier, *efficiency, control, constancy, and predictability* no longer lead to stability in our rapidly changing world. As our world and societies are changing, teaching and learning needs to also change with them. We can no longer predict and control the future, or what jobs will be needed. Instead, teaching and learning need to embrace and enable *persistence, adaptiveness, variability, and unpredictability*. To shift the mechanistic educational paradigm, students need to learn in context through all their senses and through social and socio-ecological networks that accommodate diversity. Teaching and learning need to happen beyond classrooms in school buildings and grounds, their communities, and surrounding environments.

Although 21st-century teaching and learning recognize and promote progressive educational pedagogy, particularly in teacher training institutions, until we address the controlling influence of the mechanistic paradigm, progressive teaching and learning will be compromised. Caught in a mechanistic paradigm, teachers have been trying to implement progressive pedagogy while having to respond to the undermining, controlling influence of curriculum and centralized assessments focused on separate subject disciplines. From the 19th to 21st centuries, educators such as Dewey, Steiner (Uhrmacher, 1995), Roth (1992), and Hopkins (2012) have made valiant attempts to reform education and advance progressive pedagogy, yet innovations have been sidelined with very little lasting effect. After being trained to think in bits and parts and being trained to think in terms of separate subject disciplines, from K–12 and then through teacher education, we rarely see how science, physical education, social studies, language arts, mathematics, art, and music are connected – or how working with the head, both creative and analytical, the heart, and the hands in concert with those all around us, in meaningful environmental contexts, can lead to greater understanding. From imbedded mechanistic root metaphors, our teaching, often inadvertently, maintains the status quo and entrenches the mindset of an outdated, mechanistic worldview.

The Power of Teaching and Learning Based on Ecological Principles

Sustainable education changes all that. When teaching and learning are based on the ecological principles of sustainable living systems, progressive educational teaching and learning are supported. Howard (2020) identifies experiential learning, individualized instruction, deep learning, real-world application, cooperative learning, and creating learning tasks with direct relevance to students' lives as common progressive educational approaches. He further recognizes,

> [21st Century teaching and learning] initiatives represent a moving away from traditional education approaches that are transmissive, controlling, authoritarian, and demanding of compliance in behaviour and thought. The focus on creativity and imagination, collaboration, and critical thinking is hopeful, as is the renewed emphasis on the importance of student-teacher relationships.
>
> (p. 17)

As highlighted earlier, we don't need to invent anything new: we just need to turn to the wisdom imbedded in sustainable living systems to learn how we can adapt teaching and learning to support rather than subvert the innovations we need. Applying ecological principles to teaching and learning provides us with a guiding framework and opens opportunities to further innovate and redesign teaching and learning as an adaptive, emergent process developing sustainability, innovation, and creativity. It means learning to teach and teaching using a systems approach (Meadows, 2008; Senge et al., 2012), recognizing the whole and the complex relationships between various parts, such that the whole becomes greater than the sum of its parts.

Teachers and students are also integral aspects of these dynamic, interacting systems. As such, they are empowered as equal co-creators of their learning through initiating learning opportunities that foster *feedback, cycling, adaptation,* and *emergence.*

Grounded by the roots of eco-centric systems thinking, the ecological principles are the core strength of the system, giving rise to the network structure of education and the diverse aspects of teaching and learning that support the development of sustainability, creativity, and innovation. How these are brought to fruition through each of the ecological principles will be explored below. Looking more specifically at each of the ecological principles, as the core of teaching and learning, will give greater insight into how teaching and learning will unfold; and how they can become the necessary foundation for sustainability, innovation, and creativity needed for all to thrive in the 21st century.

Interdependence

In recognizing interdependence as a guiding principle of education, teaching and learning support rather than subvert whole-body learning and systems thinking. Students learn to see themselves – as they are taught and learn in context – as an integrated, interdependent aspect of the whole of society and the natural environment through our socio-ecological interdependence. Teaching and learning need to nurture an emotional bond and love of Nature, (re)connecting through outdoor learning experiences so that ecological principles are understood innately. This may be through working in school gardens, sitting quietly in the natural environment being place-responsive, drawing or writing in a special outdoor spot, or engaging in various activities in diverse outdoor environments, learning from Nature. In this way, the whole of the student, their mind, body, and environment, are interdependently contributing to their understanding of the systems we are integrally a part of.

We now know from neuroscience that our right- and left-brain hemispheres actually work together with our emotions and whole body in relation to our environment. For example, when kids are learning a new skill or concept, they use their creative right brain to take in all the sensory information they are hearing, seeing, touching, smelling, and tasting, learning through movement of the body. All the while, their analytical left brain is categorizing and organizing that sensory information associated with past knowledge and experience and projecting future possibilities. How they feel about the experience, along with their emotional connections, are typically influenced by their environment and those they interact with; these factors will affect whether it is a positive experience that will encourage retention and further learning (Goodlad, 1984). As well, learning with others, and being in a variety of environments engages multiple senses, leading to greater connections and deeper learning experiences (Chawla & Escalante, 2007). Integrating music and the arts is key to engaging multiple senses as well as multiple ways of knowing.

By teaching and learning holistically, the interdependence between our cognitive, affective, emotional, health and well-being is honoured and strengthened. This aligns with O'Brien's powerful concept of "sustainable happiness": "Happiness that contributes to individual, *community*, or global well-being and does not exploit other people, the environment, or future generations" (O'Brien & Howard, 2020 p. 3); and a key value of "promoting the health and well-being of students, staff, the wider community, and natural environment" in Living Schools (O'Brien and Howard, 2020, p. 4). This is also supported by the First Nations Educational Steering Committee (FNESC) (n.d.), First Peoples Principles of Learning, that highlights, "Learning ultimately supports the well-being of the self, the family, the community, the land, the spirits, and the ancestors" (p. 1).

Curiosity facilitated through inquiry-based learning is key to empowering these holistic learning opportunities that are centred on experiential, place-based learning. In strengthening the interdependence between students, teachers, and the learning environment, we need to learn to trust that students want to learn by developing reciprocal relationships, empowering them to initiate and take responsibility for their learning. Questions generated by students, by the environment, and by teachers are critical to encouraging a sense of wonder and curiosity. Up until school age, kids learned naturally from everything they did and with everyone they interacted with. Up until formal schooling kids actively drive their learning. They teach themselves, in conjunction with their families and communities how to talk, how to walk, run, and how to socially interact to fit in with their communities through their own curiosity.

All of this shows we are interdependent, interconnected learners, connected to and influenced by as well as influencing the larger systems we are a part of (Chawla & Escalante, 2007; Sterling, 2010). Emphasizing whole-body learning in the environment develops systems thinking through greater understanding of socio-ecological relationships and stronger, positive environmental relationships. As humans are interconnected, influenced by, and influencing our natural environments, systems thinking and inquiry-based teaching and learning in our natural environments help strengthen our understanding and acting effectively on this principle. Recognizing the importance of interdependence between learners, between students and teachers, and between students, teachers, the human and more-than-human environments, a living systems educational framework enables teachers to facilitate and reinforce those interdependent relationships, further supporting and enhancing our innate curiosity to learn.

Just as important, subject disciplines are recognized as interdependent, each with diverse perspectives that contribute to a holistic understanding. Teaching through subject integration brings out the interrelationships between disciplines, further helping students understand and develop systems thinking, which is essential in learning to deal with complex socio-ecological issues we are facing. Climate change cannot be dealt with separately in science. It is also a significant social issue that needs the integration of scientific understanding, social perspectives, history, geography, governance, transportation, energy use, recreation, and communication as expressed through language arts, music, and the fine and performing arts. Understanding interdependence and seeing patterns that connect will open opportunities for deeper insights, innovation, and creativity, encouraging us to think in systems making connections between ideas, concepts, content, and real-world applications.

As we noted in the introduction to this chapter, in rethinking how we teach and learn we need to address how we structure and schedule teaching and learning in order to develop and support systems thinking and responsive learning communities. At the elementary level, teachers have traditionally taught all subjects making integration of subjects and place-based learning beyond classrooms easier to achieve, yet subject specialist teachers are increasingly becoming part of curriculum delivery. At the secondary level this is even more challenging, but subject specialists will need to collaborate to team-teach topics, bringing in their various perspectives and competencies to help exemplify and encourage interdisciplinary systems thinking and learning.

Community

Recognizing our interdependence, rather than seeing ourselves as isolated individuals, highlights the importance of communities and the need for positive interrelationships. This ecological principle of sustainable living systems seeks to optimize the individual as part of and contributing to the systems they are imbedded within. In this context, education and schools lose their traditional boundaries of focusing on specialization and individual achievement in age-defined classrooms isolated from their communities to become integrated into our communities. Learning becomes structured on learning communities of mixed ages, and interests and abilities based on local, personal, first-hand, contextualized learning opportunities facilitated by teachers as well as community members with the support of teachers. In order to do this, teachers need competencies to create opportunities for sharing ideas and experiences from different disciplines, places, cultures, and generations without prejudice and preconceptions; and to be able to engage with learners in ways that build positive relationships.

Typically, we designed schools to be separate from society, programming learning to happen Monday to Friday 9 am to 3 pm. But what happens all the other hours and days of the week? Is learning really separate from play and daily living? Up until school age kids learned all the time, no matter what they were doing, no matter where they were.

Once school becomes associated with learning in context, and the school doors are open to places of learning in both our human and more than human communities, at any time, classroom walls no longer seal off the community and natural world – and the sky's the limit! Teaching and learning in context, in natural environments and the community, provide a relevant context such that experiential, outdoor education becomes centrally relevant as a vehicle for learning; and Nature becomes the teacher. In this way, the community and all its resources both natural and human-made become contexts and provide opportunities for learning. Teachers will encourage learning in the community, in natural systems such as fields, forests, streams, gardens, lakes, mountains, deserts, or oceans, as well as diverse indoor learning environments.

To bring community learning to life, students will be involved on a day-to-day basis in learning at a variety of community locations with community members of all ages working side by side with children of mixed ages. The idea is to learn in the most appropriate place with an openness to learning from Nature and community members as well as fellow students. In this way, schools will be available to all learners of every age, encouraging life-long learning and community development. Focusing on the ecological principle of *community* helps teachers move beyond the classroom to use the natural, social, and built environment, as well as their own institution, as contexts and sources of learning.

While our traditional system stresses competition and *survival of the fittest*, a natural community is predominantly based on both cooperation and competition. Cooperation is essential in building and learning in communities, and in appreciating and honouring diversity. Competition, from a positive perspective, focuses on encouraging, challenging, and supporting each other to strive to be the best each can be. When combined with cooperation, the integrity of the whole community becomes greater because of the sum of its parts.

Diversity

Diversity is a key ecological principle of sustainable living systems that leads to resilience. As sustainability and learning are living processes, rather than products, encouraging a diversity of learning styles, initiatives, and approaches leads to enhanced critical and creative thinking, inductive thinking as well as deductive reasoning, empowerment, and innovation in our rapidly changing world. When we translate this principle further into how we evolve teaching and learning, our traditional notion of schooling starts to open up as students start to take part in a diversity of learning approaches, a diversity of groups in a great variety of locations so as to encourage resilience through the development of a diversity of skills, abilities, and intelligences according to diverse needs. Recognizing the value of and enabling diversity frees the teacher from trying to channel all students with diverse needs and challenges into achieving the same predetermined learning outcome in the same timeframe. Instead, all are free to learn according to their needs and interests.

In Nature all species have something special to contribute. In education we need to encourage diverse learning opportunities so students can develop and contribute their special gifts. As diversity also highlights the importance of using and developing multiple intelligences to learn, making diversity a core principle opens opportunities to make learning relevant for everyone. Diversity encourages teaching that enables an ever-changing learning process that is responsive to changing individuals in their local contexts and provides a diversity of learning options and resources to maximize learning potential.

To realize this, teachers and learners take on a variety of roles from learner to facilitator or mentor, in a variety of groupings working collaboratively through mutual understanding, tolerance, and cooperation while respecting diversity. Not everyone will learn the same thing, in the same way, at the same time, so we need to honour and teach with this diversity in mind. As with the ecological principle of *community*, *diversity* supports the need to learning in diverse environments beyond the classroom. This helps teachers as well as students use all their senses to assimilate new learning to adapt and change. This is what we need. A way of teaching and learning that encourages diverse transformations. When learning has personal relevance, when students are involved in the decision-making process, and learning occurs with others in context through a diversity of senses, it becomes more dynamic, meaningful, motivating, and empowering.

With open, adaptive teaching, diversity is not only honoured but also encouraged. The arts, hands-on activities, and critical thinking that encourage imagination, innovative ideas, and empowerment are emphasized. Diverse forms of knowledge (tacit, theoretical, technical, scientific, traditional ecological, cultural, poetic, spiritual, bodily) are all to be recognized and encouraged within a systemic context and enabled through innovative teaching and learning approaches. In highlighting the importance of diversity through the concept of pluriversalism, Perry (2023) recognizes,

> Pluriversalism connects strongly with the UN Sustainable Development Goals in that it recogniz[es] the diversity of people's views on planetary well-being and their skills in protecting it. Embracing a pluriversal approach has the potential to push back against inequalities and galvanize educators into adopting pedagogies that champion Education for Sustainable Development (ESD).
>
> (para 2)

Incorporating Traditional Ecological Knowledge, Indigenous knowledge and ways of learning is essential in decolonizing education and incorporating pluriversalism in teaching and learning by increasing Indigenous perspectives in lesson plans and learning activities. In this way, Indigenous worldviews are given the value and respect they deserve instead of being seen as a token activity or add-on. Passing down traditional knowledge to younger generations by developing positive relationships with community members and elders in Indigenous communities, and working directly with local elders and knowledge keepers, provides a more pluriversal understanding of sustainability, helps non-indigenous students expand their worldviews and understanding, and allows Indigenous students to reclaim their heritage (FNESC, n.d.).

Recognizing the need to incorporate these numerous dimensions of diversity in teaching and learning can lead to feelings of stress and anxiety when administrators and teachers are seeing this through the lens of the industrial schooling system. Teachers are expected to teach to diversity in their classrooms and there is significant pressure on both teachers and administrators to do so from parents, school districts and Ministries of Education. But our industrial schooling system is not designed to enable and support teachers to do so. This leads to stress and frustrations of administrators, teachers, students, and parents given a lack of support, lack of funding, and expectations for student success and graduation rates in a system inherently designed for conformity rather than diversity. It is essential to recognize teachers cannot effectively incorporate diversity on their own. This underscores the need to develop a holistic view of education recognizing the inherent interdependence of the educational paradigm, organizational structure, administration, leadership, and community with teaching and learning. Administrators, teachers, and learners need us to transition to an eco-centric approach to education so as to embrace and support diversity right from its organizational structure through to its policies, funding priorities and administration support. In this way, diversity can expand our thinking and encourage and enhance student diversity, biodiversity, and cultural diversity, leading to resilient learners who can adapt and thrive in developing resilient socio-ecological relationships.

Cycling and Feedback

Rather than being a linear, machine-like system based on a top-down, plan-and-impose model, teaching within a living-system framework looks more like a living, dynamic, responsive system of *feedback loops* and iterative thinking (Senge et al., 2012). The ecological principle of *cycling* is inherent in this educational paradigm at various levels. At the larger curricular design level, the fundamental concepts and big ideas are developed and reinforced through a spiral teaching approach based on needs and interests of students and their communities as they develop over time. At a personal, yet collaborative, level, as students cycle through various topics at multiple age levels, they have opportunities to explore new topics or revisit topics to allow for deeper understandings, seeing various concepts in new, diverse ways. Throughout, the emphasis is on *feedback* to encourage *adaptation* and *emergence* in teaching and learning. Teachers model a partnership rather than dominator role in the classroom focusing on peaceful, interpersonal skills rather than domination, control, and imposition so as to support child-responsive teaching and learning.

The Adaptive Learning Cycle

The process of learning is also cyclical. This living systems framework enables each student's transformative potential through an active learning cycle of direct experience, critical reflection, abstract conceptualization, and active experimentation in developing and applying new concepts and

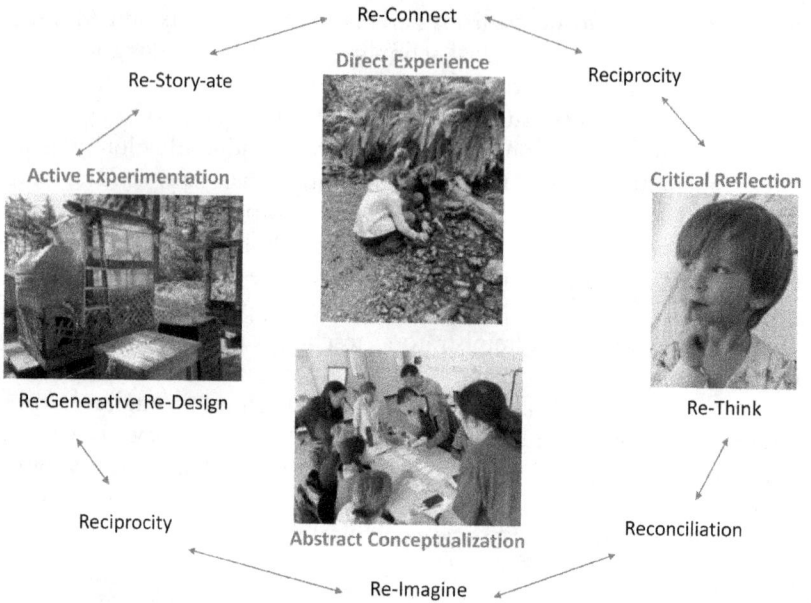

Figure 7.1 Active learning cycle for transformative inquiry-based learning. Designed by the author adapted from Kolb (1984).

understanding (Institute for Experiential Learning, 2020; Kolb, 1984). Kolb's active learning cycle is an excellent basis to incorporate new Rs for this, modelled in Figure 7.1.

Through reconnecting in real world contexts, students gain direct experiential learning in diverse environments. This helps students and teachers understand interdependence and *reciprocity*, to *rethink* our socio-ecological understandings. Such new awareness leads to further learning for *reconciliation* between cultures, recognizing pluriversalism and the importance of respectful communities, as well as reconciling our socio-ecological relationships so all thrive. *Re-imagining* and *reciprocity* lead to *adaptation* based on *feedback* so that *regenerative redesigns* can emerge to re-story-ate in transforming to sustainability. This is supported by the FNESC (n.d.) in identifying a key principle of First People's learning: "Learning is holistic, reflexive, reflective, experiential, and relational (focused on connectedness, on reciprocal relationships, and a sense of place)" (p. 1).

Adaptation and Emergence

As sustainable living systems develop stability and resilience from *persistence, adaptiveness, variability,* and *unpredictability* (Holling et al., 2002) teaching and learning need to enable the development of these characteristics by embracing adaptation and emergence as core principles to help students become resilient and thrive in our dynamically changing societies. In helping prepare students to engage effectively in our complex world and the challenges in developing the mindset and skills in transitioning to a sustainable society, we need to be able to expect the unexpected, seeing unpredictable occurrences as positive learning opportunities. Engaging with and embracing variation, change, adaptiveness, and persistence is to embrace transformative learning. These qualities that lead to life-long learning are not overtly taught but are developed through teaching that is open, adaptive, and emergent rather than focused on predetermined learning outcomes in sequenced units and lessons in separate subject disciplines. To do this, teacher training institutes need to ensure teachers develop the competencies to be a facilitator and participant in the learning process; a critically reflective practitioner; inspire creativity and innovation; and connect the learner to their local and global spheres of influence (UNECE, 2012).

When we apply this living-system model, we recognize that change and learning evolve from the ground-up: through transformation rather than transmission. Using this approach, "teaching" takes on a whole new meaning. It is more about mentoring, challenging, encouraging, and supporting a diversity of ways to learn. Teaching and learning need to adapt and emerge, as the students become involved in constructing their knowledge, understanding, and evaluating their learning as they engage with others in diverse contexts. Creativity is emphasized throughout with creative, open-ended inquiry that enables innovation and adaptation. The arts, hands-on activities, and critical thinking help prioritize imagination, innovative ideas, and empowerment to enable the development of transformative learning opportunities.

As our world is continually adapting, emerging, and changing, you can never "know" for certain or attain total knowledge about any particular subject – as traditional schooling has led us to believe. Knowledge is not 100% certain and fixed in time. Therefore, we shouldn't teach as if it is. Instead of expecting students to completely *master* a particular topic or subject, it is more realistic and helpful to have students and teachers learn deeply and understand by *engaging* with various topics so they can bring in any number of interconnecting threads of thought in developing a growing systemic understanding while being open to changes – as well as further innovative ideas. In this way students become comfortable with change and adapting as we continually learn more.

Through adaptation, teaching also allows community-based learning to emerge as it becomes locally relevant. This process of *adaptation* and *emergence* mimics innovation in sustainable living systems, enabling innovations and emergent learning rather than trying to implement predetermined activities to elicit expected responses. In being open, adaptive, and emergent through teaching based on true inquiry learning, innovation and creativity rise to the forefront, empowering learners and teachers to explore the unknown and respond effectively with inner motivations in developing learning capabilities. As such, this teaching/learning approach develops sustainability as a frame of mind (Bonnett, 2002) – as an emergent way of seeing and interacting systemically in the world.

Assessment

How we evaluate teaching and learning is key to breaking free of mechanistic teaching and developing systems thinking through *feedback* to enable *adaptation* and *emergence*. No matter how innovative a teacher wants to be, if education is being controlled to, or the teacher perceives they must adhere to, a predetermined set of standardized learning outcomes due to a hidden mechanistic paradigm, the teaching and learning will reflect that directive (Ireland, 2007). To provide more innovative empowering education, teachers need to be free from having to teach and be accountable to predetermined learning outcomes and standardized tests. Alfie Kohn (2002) has undertaken significant research related to the negative impacts standardized tests and grading have on learning. He summarizes these findings:

> To read the available research on grading is to notice three robust findings: students who are given grades, or for whom grades are made particularly salient, tend to (1) display less interest in what they are doing, (2) fare worse on meaningful measures of learning, and (3) avoid more difficult tasks when given the opportunity – as compared with those in a nongraded comparison group. Whether we are concerned about love of learning, quality of thinking, of preference for challenge, students lucky enough to attend schools that do not give letter or number grades fare better.

Yet teachers and administrators routinely perceive our mechanistic approach to teaching, learning and assessment needs to continue as the K-12 educational system is at the mercy of post-secondary requirements. As a practicing administrator recently summarized, "Percentages and ranking drive the education world, not how people think, create or experience learning."

This linear concept of education based on the industrial root metaphor of *competition and survival of the fittest* is causing untold stress and anxiety for students, parents, teachers, and administrators. It is deeply imbedded in parents' minds as they want their child to "succeed" with top marks to get into the "best" schools to secure future success. This was starkly exemplified in a recent conversation with a mother of a one-year-old, who was already stressed about finding the best day care that would get her daughter into the best feeder elementary school that led to the top-ranked secondary school, so she could eventually be accepted in the top university of her choice. Another example of this problematic root metaphor, came from China: when working there in 2019, parents I know in Tianjin were moving to more expensive, far less affordable districts so their child could get into top-ranked schools from K-12. These parents would get daily updates as to how their child ranked in the class that day, and during parent-teacher meetings, parents had to sit in rank order depending on how well their child is performing. These may sound like extreme examples but they are not uncommon for many. In this competitive industrial system of assessment and ranking, students as well as parents are experiencing significant stress as the emphasis shifts from learning to performing on tests.

This doesn't need to be the case. We are recognizing at the post-secondary level that having the ability and self-motivation to learn, collaborate, think critically and creatively, adapt, think systemically, and apply learning in real-world contexts is essential in our changing world. As such, more and more post-secondary institutions are considering portfolios, references, and interviews as admission requirements, and students who wish to follow a particular post-secondary program can study separately for and take an entrance exam, when necessary. Oxford University, for example, in their admission interviews for medical school, are not asking what you know, but are more interested in how you think and your ability to use your knowledge and experience to be innovative in solving problems. Our world has changed. The unsustainable industrial schooling system, with its embedded mechanistic root metaphors and assessment practices is not going to help our children thrive in our world where successful careers and quality of life will depend on adaptability, innovation, creativity, and collaboration rather than competitive conformity. Now more than ever, we need a society of learners, able to thrive in developing a sustainable society through innovation and creativity. This speaks to the need to rethink the purpose of education in what and how and what we are assessing. When we focus on learning and the need to develop a sustainable society as our priority, assessment necessarily changes.

To encourage teachers and students to develop and understand we are part of a dynamic *interdependent*, constantly adapting world, develop and work effectively in both human and more than human communities, develop and respect *diversity* of their skills and talents, celebrate *diversity* in others, and recognize the importance of, and abilities to engage in *cycling*, *feedback*, *adaptation*, and *emergence* in learning, evaluation will shift from standardized testing to authentic assessment that is developmental and process-oriented; recognizing where teachers, teacher trainees, and students are at so as to inform continuous learning cycles. Rather than evaluate students' ability to perform to predetermined levels in narrowly defined learning outcomes or performance competencies, students and teachers collaboratively evaluate their ability to think divergently, their ability to innovate, see interconnections, as well as their abilities to learn, adapt, and think in terms of systems.

As we saw in Chapter 4 when discussing evaluation, teachers need to become reflective practitioners, developing and sharing their learning in professional learning communities within, across, and between scales: with their students, parents, fellow teachers, and administrators. For this model to work it is essential the system itself is designed to incorporate innovative opportunities in teaching, teacher training and learning throughout K–12 that are "safe-to-fail,", so that creativity, innovation, and learning are encouraged and supported for both successes and perceived "failures". In this way, not getting a result or an answer that works is not seen as a failure, but as a learning opportunity to innovate and try something else.

The First People's Principles of Learning highlight the importance of being responsive to *feedback* and that learning is *interdependent* and reciprocal through their third principle that recognizes, "Learning involves recognizing the consequences of one's actions" (n.d., p. 1). Students as well as teachers are actively encouraged to own their learning by recognizing challenges as learning opportunities and setting personal goals through sharing their continuous self-assessments. This helps teachers, teacher-trainees, and students understand how to communicate learning and think critically about the process of what and how they've learned.

Given that teaching and learning are co-dependent, evaluation necessarily applies to assessments of teaching effectiveness as well as student learning. The students' and parents' perspectives are essential in creating teaching and learning as a collaborative learning cycle based on student, parent, and teacher feedback. Through a constant cycle of reciprocal *feedback*, parents, students and teachers become partners in reflective practice. This will develop further resilience as students, parents, teachers, and the educational system support being able to adapt and emerge as conditions around us change. Rather than being seen as a threat, our unpredictable, ever-evolving future will be seen as a positive opportunity for continuous learning.

Summative evaluation for teacher trainees, teachers, and students will have a role but only in the context of adapting and generating a new learning cycle. This moves beyond the evaluation *of* learning to emphasize developmental evaluation *as* and *for* learning. This focus on the process of learning and adapting helps us put effective pedagogical approaches front and centre where both teaching and learning are continuous, *emergent* properties. As such, our evaluation metrics must also continually *adapt* as teaching and learning continue to *adapt* and *emerge*.

Practical Applications

The Green School Scotland

As discussed in the previous chapter, The Green School in Scotland evolved using the Eco-Centric Curriculum Framework so teaching and learning were topic-based, designed to bring out the ecological principles as the core of the curriculum, with various subjects providing diverse lenses and insights in developing systems thinking through experiential, place-based inquiry learning. As noted earlier, students from ages 5 to 12 attended the school in multi-age groupings so the topic-based curricular design provided an open structure for the teacher to respond to diverse learning needs and interests. Teaching and learning took place outdoor in Nature and human communities, and in the classroom when shelter, a kitchen, or tables for expressing learning was needed.

Once curriculum topics were identified, based on the students' and teachers' interests, various teaching and learning activities unfolded organically from the initial curriculum design brainstorming session. As noted earlier, numerous *who, what, when, where, why and how* questions came up during curriculum development and these were then used to help inspire inquiry-based learning opportunities. Being aware of competencies and how various subject discipline lenses could be used to explore these questions systemically, the teacher often provided learning opportunities that were interdisciplinary and experiential, engaging the head, heart, and hands. Facilitating learning through the outdoors provided the basis for this holistic learning through the active learning cycle (Figure 7.1), enabling Nature to be the teacher, often giving rise to unexpected insights and new learning opportunities.

Each day provided both structure and flexibility in teaching and learning so that the teacher and students could effectively embrace variation, change, adaptiveness, and persistence. To start each day, the teacher and students gathered in a circle to reconnect by sharing news and ideas for the day's learning activities. These conversations could change what would happen, extend, or reinforce previous plans. Then we would typically head outside with backpacks full of learning materials to help reconnect with the more-than-human world through experiential learning activities to initiate an active learning cycle.

Explorations to reconnect through direct experience typically started with heading up to our favourite look-out to read the weather, observe changes from a bigger perspective, and then move to our special sit spots to spend time reflecting and reconnecting on a personal level immersed as part of a very small setting, becoming place-responsive. When the students first found their special spots, it took time to settle in and feel comfortable. But after spending daily time in those places, students felt part of their environment, often observing ongoing activities of insects and animals they shared their spaces with, as the animals became comfortable with their presence. This was a time of quiet reflection, but at times students kept a journal or drew to help capture their insights, critical reflections, and growing awareness of our interdependence. No matter what topic was the focus of the day, innumerable transformative insights and experiences came from Nature being their teacher.

Other opportunities to learn from Nature involved more active direct experiences. For example, in exploring Food as the curriculum topic, students were outside in natural and human communities using multiple senses to explore elements, plants, animals, and people, bringing out diverse ways of learning and knowing based on the food-related inquiry questions that motivated them. As an example, these explorations led to finding food various animals eat in the local forest while playing "The Survival Game", where students became an omnivore, herbivore, or carnivore, needing to locate food and water sources or catch an animal by tagging them and taking one of their life tags to prove they've survived. What seems like a simple game leads to significant insights into the life of animals through lived experience of the importance of getting food, using all one's senses to be alert to those that may be trying to eat them, while meeting their own need to find food and water, camouflage and hiding, as well as fitness to escape, leading to understanding the ecological principles of *interdependence, community, diversity, energy flows and cycling, feedback, adaptation, and emergence.*

This experience and the critical reflections that followed led to abstract understanding of the importance of biodiversity and effective socio-ecological relationships, particularly when a human came into the game introducing pollution that travelled up the food chain through a food or water source, or fire, and only had to see an animal to take its life. In some cases, their experiences also sparked student-initiated learning activities in reading and research on various animal adaptations, creative writing from an animal's perspective, and research into traditional nomadic, vegan, and vegetarian diets. On a few occasions students became particularly interested in a particular animal based on their experiences and developed an independent learning project to follow up with. This particular learning cycle developed into rethinking and re-imagining food options by researching and creating a vegan lunch from local food sources. Being so open-ended students from ages 5 to 12 could all participate in and *interdependent* learning *community*, developing learning according to their interests and developmental levels.

Being aware of the need to develop diverse competencies, the teacher would facilitate *diversity* in activities to further learning in the topic.

Through inquiry questions that could lead to developing mathematics, or communications competencies, for example, the teacher took on the role of guiding and mentoring. Simply asking, "I wonder how many types of seeds there are around us," "I wonder how much water a tomato plant needs to make tomatoes?", or "I wonder where we can find out the best type of vegetables to grow in our area?" open opportunities for developing competencies in ecological literacy, communication and media literacy, information technology, critical thinking, problem-solving, systems thinking, creativity, authentic real-world learning, health and well-being, adaptability, and lifelong learning. Specific skill development in mathematics at various levels, as well as communication skills in language arts were naturally integrated.

Developing competencies in mathematics came from the teacher facilitating opportunities to learn and apply mathematical concepts to explore various topics and express their learning. Initial experiential explorations often led to wanting to develop further mathematical competencies. Being a multi-age class, younger students often saw older doing things they were not yet able to, younger students might ask questions to inspire older students, and the teacher would ask inquiry questions that would stretch all student's abilities, often leading to creative innovative problem-solving. A good example of this stemmed from younger students wanting to play "Pooh Sticks" from reading about it in a *Winnie the Pooh* book, from looking at animals that live in a forest. To play this game students throw sticks in the river from a bridge and see whose comes out first on the other side, with younger and older students keen to play. This led to timing sticks, and then questions regarding stream flow at different places across the river, the impact of river depth, obstacles, and current. The afternoon morphed into developing some mathematical competencies needed to answer these many questions.

In knowing the students' levels of competency, the teacher would ask various mathematically based inquiry questions to further explore the topic and motivate students. When students needed to learn new concepts and skills to further explore the river, the teacher mentored students, in groups or individually, in learning how they might find their answers, according to their various levels. Such explorations involved learning through number operations such as adding, subtracting, multiplying and dividing, trigonometry, or algebra; or expressing their learning through pattern recognition, geometry, measuring, and data representations.

The next day the students brought stop watches, measuring sticks, tape measures, drawing materials to the river to map the river profile in various places and measure stream flow in various locations, as well as the book *Winnie the Pooh*. This led to untold calculations, graphing, and applying new skills in creating a stream profile, correlated with graphs of stream flows. Based on this critical reflection and abstract understanding, groups of students strategized how to apply their learning and work with the river, experimenting anew in a more intentional game of Pooh Sticks. After all the activities, students had lunch by the river, watched how a diversity of species lived in and around the river, and used it for seed dispersal. Students then took time to find a special spot for quiet reflection and reading, with one boy having thought ahead to bring a folding chair so he could sit in the middle of the river to contemplate as Pooh had ☺.

By connecting to their interests in the river from direct experiences, problem-solving was real rather than contrived. Rather than starting from decontextualized mathematic concepts and procedures, and then trying to apply them to a fictitious problem that lacks meaning to the student, learning became meaningful and contextually relevant, highlighting *the whole being greater than the sum of its parts*.

In developing competencies in language arts learning was also contextually relevant, integrated into the topic, and flowing from experiential activities. In the early grades reading was developed from reading their own books, which they illustrated and may have need to write their words with the assistance of the teacher. For all levels, reading material related to the topic was accessed and often read outside in relevant contexts. In developing writing skills students often wrote outside where they were motivated to record their thoughts, discoveries, and creative expressions. Learning standardized spelling came from words they needed to spell in their own writing. This made writing purposeful and much easier, with students happily engaged in a *diversity* of writing activities depending on their interests and purpose. The power of writing in context was exemplified by a 12-year-old student who had experienced traditional decontextualized classroom education before coming to The Green School. One day, while the class was up on a high hill, involved in various explorations, she took herself off to the side and became immersed in writing. When checking in with her, she was so excited saying, "I've always hated writing because in a classroom my ideas are blocked by walls - but up here my ideas can flow so easily! I love writing now!"

Throughout the week time was flexible to allow for these unpredictable teachable moments, and personal as well as small group self-directed learning opportunities. This is where the teacher became a facilitator, mentor, and at times a co-learner, *adapting* to a reciprocal relationship in order to facilitate adaptive learning opportunities. In this way learning was truly organic with the teacher developing and supporting the conditions for learning to thrive.

Assessment

Assessment at The Green School was developmental, based on continuous *feedback* to enable *adaptation* and *emergence*. Students set personal learning goals and asked for feedback from the teacher as well as co-learners.

The teacher frequently had discussions with students individually and in groups to empower the students to own, *adapt* and *emerge* their learning by asking how learning was developing, what challenges they faced, and then discussed ways they could further their learning. At times these discussions flagged the need to develop new competencies, or at other times, cycle back to strengthen others. By giving students ownership over their learning through setting their own goals, self-assessments and seeking constructive feedback, students were more focused on the process of learning and became self-motivated in setting further personal learning goals. This was exemplified well when a government inspector came to evaluate The Green School. When I needed time to speak with him, I asked the students what they would like to do on their own. While many continued to work on their group or individual projects, researching or representing their learning through writing, graphing, art, and dramatizations, one boy said he wanted to work on memorizing his times tables to help him be quicker in his calculations. The Inspector was astonished saying in all his years as a teacher and inspector he had never heard a child voluntarily choose to learn times tables.

Feedback was also central to developing community-based transformative learning competencies. As learning is an interdependent process involving the teacher, the other students, and the communities they interact and learn with, continuous feedback helped develop students' abilities to develop, and interact effectively in, collaborative, respectful relationships. Through authentic real-world learning, and responsive, collaborative feedback, students were encouraged to expand their abilities to expect the unexpected, see unpredictable occurrences as true and positive learning opportunities, and engage with and embrace variation, change, adaptiveness, and persistence in developing transformative learning competencies. Developing these competencies was seen as an exciting aspect of learning and students naturally included them in their personal development goals, often reflecting on their progress in their personal learning journals.

Focusing on assessment as and for learning, individually and collectively, highlighted the ecological principle of *feedback* to enable *adaptation* and *emergence*, a central component of how the active learning cycle unfolded. Following direct experience, critical reflection led to deeper abstract understanding in collaborative learning with others, and further active experimentation in developing and applying new concepts and understandings. This fostered a love of learning, recognizing the value of collaboration, and the ability to *adapt* and take on leadership roles in applying solutions to *emergent* challenges. Three of the students I have stayed in touch with have taken this foundational educational approach with them, exemplifying their self-directed, life-long learning approach, and ability to develop solutions through very successful careers in furthering a sustainable society in environmental, economic, and social sectors. One co-founded Materiom.org, growing the regenerative materials economy with open data and AI. Materiom offers a database of recipes for making biodegradable materials using abundant natural building blocks, and helps scientists and entrepreneurs to accelerate research and development to move from fossil based to regenerative materials. The second is enabling solutions as a Vice President in the financial banking sector to help support technology companies innovate and respond to the changing needs of society. The third is a medical doctor and researcher working on cutting-edge genome research to find treatments for diseases such as amyotrophic lateral sclerosis (ALS) and endometriosis. By practising and internalizing this continuous cycle of assessment for transformative learning, the students developed competencies to become lifelong, self-directed learners with sustainability as a frame of mind.

A Bioregional Approach

The bioregional school in my case study research taught the traditional provincial curriculum, but through a bioregional approach, also modelling eco-centric teaching and learning. As the provincial curriculum provided schools with the freedom to teach the mechanistic provincial curriculum in a variety of ways, it was encouraging to see the bioregional school exemplify much of the same approaches to teaching and learning as The Green School, in consciously creating learning opportunities based on an eco-centric paradigm. In bringing to life their bioregional curriculum, the bioregional school identified independent learning, democratic community of learners, elders and kids, apprenticing, and child-directed learning as central approaches to learning. In creating a vibrant learning environment, the staff and volunteers' interest and enthusiasm sparked the students' interests and modelled life-long learning.

View of Learning

The view of learning at the bioregional school is very systemic. Children, parents, teachers, and mentors are encouraged to learn together in a learning environment that is safe and inclusive of all levels of learning. The school also emphasizes that education needs to focus on development of whole people where physical, intellectual, emotional, & spiritual growth are fostered, in accordance with an ecological view that sees the learner as a whole person with a full range of needs and capacities. As the school manual states,

> Individuals have unique interests, feelings and learning styles. Successful learning is built on the learner's prior experience and stems from their interests. All learning experiences should be enjoyable and challenging. Students are not pushed to learn before they are ready and are not held back from learning, based on age or grade.

The school also stresses that learning is an integrated process. In doing so it emphasizes ecological metaphors of holism and systemic thinking. This accords with Bowers (1995), Orr (1994), and Sterling (2001) who suggest that the curriculum should encourage more trans-disciplinary domains of interest rather than disciplines and a defence of discipline boundaries if it is to be characterized as ecological rather than mechanistic. It also resonates with the eco-centric paradigm by incorporating transformative education and an ultimate concern with wisdom. In emphasizing the ecological principles of *interdependence* and *the whole is greater than the sum of its parts,* the school manual states,

> Problem solving, critical thinking and cooperation are life skills that transcend all subject areas. More importantly than specialized knowledge, students need to know how to learn and how to make connections amongst the things that they know. Integrated studies are well suited to developing a mature understanding of the world and an ability to think clearly to affect social change.
>
> Learning tasks need to be authentic and holistic. Breaking things down into component parts is not the only way to understand the world. Children tend to think in holistic ways and enjoy challenges that are real and in context. These challenges rarely have a single step or a unique answer. Understanding the way things are connected into whole systems is as important as knowing about the individual components.

The school's child-directed learning model places the child's natural curiosity and innate need to learn at the centre. The view of learning needs to be meaningful first and this meaning is constructed and negotiated by the student and teachers. As one parent noted:

> Education should spark within children the desire and love of learning. From that anything can follow. The true majesty of learning has to come first so they want to learn, love to learn, interested to learn. The purpose is to spark that and then empower the children to learn themselves.

Various learning skills such as communicating, inquiring, problem-solving, taking action, understanding, and seeking new perspectives are developed through a holistic approach with learning occurring naturally in context.

In addressing learning concerns in the specific subject areas of reading, writing, and mathematics, the school manual also resonates with an ecological view in that it focuses on child-centred learning that follows a natural process and emphasizes local, personal, applied, and first-hand knowledge. The manual states reading, writing, and math are skills that come quite naturally to children once they are ready, emphasizing learning in context through authentic experiences. In this way children are encouraged to read by reading and to write by writing; before doing pencil and paper arithmetic, that same arithmetic is experienced and understood with concrete objects. The manual further clarifies,

> When children explore math authentically, they enjoy it and are proud to solve difficult problems. They show real understanding of the concepts within a problem and can use those same concepts in a variety of different contexts. They notice math in the real world and share with joy the patterns they find.

With the self-directed approach to learning, students are very much involved. They have the ultimate decision on whether they will learn or not, and what they want to learn. If they choose not to engage with a workshop a teacher plans, they can work on an individual learning project of their choosing. Students see their teachers as advisors who advise them on their education, help them negotiate independent learning options that meet their needs and abilities if they choose not to participate in a workshop, develop projects of interest, and develop independent learning skills, rather than push them into doing learning activities. As not all students had independent learning skills, this was an area the teachers recognized they needed to help students develop, especially if they were transferred into the school from a mechanistic educational system of learning. The Headteacher summarized,

> The aim is to develop a planning framework that is flexible so it can incorporate organic development of learning and empowerment. There is individual choice to opt out of a workshop by negotiating an educational option that meets their needs and works for others.

View of Teaching

A job description for a full-time teaching position showed they were looking for a teacher with a strong bioregional, ecological background and approach as well as someone who was willing to be mentored in the philosophy of the school. In line with the UNECE (2012) competencies for educators, this teacher is expected to not only teach sustainability (in this case through bioregionalism) but also model it in his or her own personal behaviours. Under skills and knowledge that would be considered an asset, the school identified unusual areas that relate to their bioregional philosophy. Worth noting are the native Coast Salish language, First Nations perspectives, wilderness skills (tracking, awareness, survival), organic farming, ecopsychology, ecofeminism, social ecology, and deep ecology. Obviously, the school sees the importance of the teacher having a strong ecological basis as an essential component of the success of their approach. This also suggests a resonance with some of the literature that indicates these philosophies contribute to an ecological worldview (Gough & Whitehouse, 2003; Naess, 1989; Sauvé, 2005; Sessions, 1995; Suzuki and Knudtson, 1993).

The bioregional school also required a teacher who could take initiative and develop curriculum as the curriculum is still developing and is not totally laid out. In support of this curriculum development, an experienced teacher would mentor him/her. The teacher they hired saw value in this mentoring approach, as she felt new teachers don't often get that type of support in the formal government school system. Teachers also being learners is a characteristic of an ecological view in recognizing all need *feedback* to *adapt* and *emerge* in contributing to a resilient system.

In discussing the role of the teacher, the Principal referred to the more sensitive role the teachers need to take in guiding self-directed learning, "Rather than teaching them, it is more guiding and facilitating their growth in those areas." When asked more specifically about the most important things to teach and why, she replied, "We have to teach children how to find information rather than the information itself. I don't like the words teach and teacher." This approach to teaching accords well with Foster's (2001) emphasis on developing a learning society and learning mindset.

Teaching Methods

The Bioregional Manual outlines a teaching approach to encourage child-directed learning,

> Child-directed methods are the most powerful teaching techniques. Activities that allow for exploration and experience allow children to discover skills and knowledge for themselves. Successful learning activities encourage children to take risks and experiment. Risk-taking is basic to learning new skills and experimentation is basic to gaining knowledge and understanding.

Recognizing children are innate learners with this natural capacity to learn, teachers are expected to nourish and encourage this drive to learn by organizing the space and provide situations to facilitate natural learning. In addition, teachers provide group activities that they feel will add to the experiences of the children. Even so, individual children can take from any activity only what is right for them at the particular time they are doing it. Students are always encouraged to pursue individual projects and learning goals, and they may choose to do so instead of attending the group lessons offered.

Guiding and facilitating is also the approach teachers take to enable experiential learning in Nature in developing a deeper connection with their natural environments. In responding to questions about how they teach this dimension of ecological intelligence, a teacher referred to the Earth as the teacher,

> We teach this aspect in subtle ways. We do not formally say we are going to learn about the spiritual dimension of bioregional education. There are different techniques that I do with them. One of them is something called sit spots or secret spots where they spend time sitting and observing on their own in nature. The kids love this and ask to do it often - even one student who says he hates nature. That's why I never consider myself a teacher- the Earth is the teacher. That is where the more spiritual dimensions come in. In subtle ways, through various activities you introduce those things.

The bioregional approach, being contextually based, involved teaching and learning outdoors in developing this strong sense of *interdependence* as well as *diversity* and *community*. One of the teachers felt the bioregional focus of the curriculum naturally brings out diversity in learning through all our senses,

> This reading the landscape comes from a very direct connection not just from reading something in a book but actually going out there and being in nature. So a lot of the fieldtrips we go out and have a lot of experiences in nature, opening our awareness and developing our senses, being able to hear more and see more, just basically live more fully as a human realizing our potential.

Facilitating experiential learning through diverse teaching methods was also exemplified through indoor learning experiences. The founder of the school, and Headteacher, explained how teachers not only guide and facilitate, in light of there being student choice, but also how they teach at a conceptual level,

> Students have the choice but it is the teacher's job to make it interesting so they will want to participate, others are usually listening. I try to teach knowledge outcomes at a conceptual level so there is some understanding there as to why things happen a certain way. Remembering and recalling information is based on their own interests, and is not a requirement.

To enable inquiry-based experiential learning, exploration and creativity were emphasized. A volunteer teacher recognized the value of art in facilitating an ecological understanding. "Ecology is taught in a holistic perspective through art. I'm focusing on the creative side of the environment rather than a left-brained lesson focus." In highlighting the power of experiential learning, students were learning math through woodworking. With this emphasis on experiential learning that is reflexive to the learners' interests and needs, the school's teaching approach is meaningful first with a strong sense of emergence in the learning environment.

The organic, self-directed nature of classroom learning developed when the focus was on guiding rather than didactic teaching. Teachers helped students explore materials they gather and bring together or go out and explore. They often piqued students' interest by being a co-learner saying, "Check this out. I think you'll find this interesting." or explore anything they are interested in exploring. In multi-age groupings they used one-to-one moments with students with a diversity of multi-age grouping to explore anything they wanted to explore.

This approach to teaching and learning reinforces the ecological paradigm as it emphasizes transformation and an integrative view where teachers are reflective practitioners and change agents as well as learners. It encourages *adaptation* and *emergence* through experiential learning, critical and creative inquiry, and employs *diversity* in teaching and learning. Furthermore, all those concerned understand learning to be reflexive and iterative with meaning being constructed and negotiated. These perspectives are consistent with what Sterling (2001) indicates as ecological as well as with what Bonnett (2002), Gough (2002), and Rauch (2002) argue as essential aspects of education for sustainability.

Planning

Being so child-centred, the teacher obviously needs to respond to student interests, but to ensure this happens, the teachers have thought carefully about how they prepare for learning. This was apparent when the Headteacher clarified the teacher's approach to planning,

> Planning depends on your goals and your time. We need freethinkers who initiate their own learning and develop new ideas. That's one of the reasons why we don't set ourselves up with unit plans and lesson plans. Because that limits what you talk about. I don't believe kids learn in step, by step, by step, but more a bit here, a bit here, a bit there.

This is best exemplified in their approach to fieldtrips. The Headteacher emphasized, "Fieldtrips don't involve worksheets and learning outcomes. We let them explore the environment in their own way." This emphasizes an *emergent* curriculum based on the individual learner rather than on specific grade levels and classes,

> Our program differs from many other educational programs in that the daily curriculum for each child emerges from that child's interests and experience. The learning goals are recorded once they are met; they are not always planned in advance. Children are encouraged individually to progress in their learning, but even during whole group activities, it is each individual who progresses, not the group as a whole.
>
> (School Manual, p. 3)

In providing alternative learning options, the teachers typically plan what and how specific content is taught, based on perceived students' needs, or how to provide diverse learning experiences, although parents and students can be involved and have input. Often co-operative goal setting with students affects what is taught. Planning can also involve bringing in a volunteer to teach art, deciding where their Friday fieldtrip will be or strengthening experiential learning opportunities related to diversity in the community by including mentally challenged adults as part of the art lessons and going to cultural events when it they can.

Teachers had a general outline of workshops or fieldtrips they were leading each day but purposely left time and opportunity for unexpected learning activities that were student generated. Typically, teachers provided three afternoon workshop options and core supervision for those who want an individual choice. When student suggestions did come up, teachers immediately followed them up either individually with that particular student by supporting a personal investigation or by introducing the idea to the group in case others found it interesting. Throughout the day the learning atmosphere is casual and relaxed with students, teachers and volunteers being respected in whatever opinions they express. Consistently, students are asked to do things rather than told to. They were given choices and those choices were respected as long as they were meaningful learning activities and did not interfere with other's learning.

Beneath the empowerment to make choices was an expectation that students would be responsible for their choices. If a student behaved in an irresponsible manner, it was discussed with the student so they understood the expectation, and then they were given the time and space needed to cooperate effectively or to work on their own if need be.

Learning Environment

Given this belief in sparking student interest so that learning is meaningful and full of holistic learning opportunities, the learning environment becomes an important aspect to be considered. The learning environment at the bioregional school was an exciting, positive context for both classes. Both older and younger students had a core-learning environment that catered to quiet as well as active learning experiences, and individual as well as multi-age group dynamics. There was an extensive library as well as an excellent variety of teaching resources. The School Manual recognizes the use of space is different from what one might expect in a traditional classroom. "Our rooms look less like traditional classrooms and more like eclectic living areas. We try to create spaces where interesting learning activities are encouraged by the space itself." The main classroom reflected this. Rather than desks there were worktables in one area; a piano and computer in another; comfortable couches for reading and discussion; a variety of science-related experiments and objects on the side; an eating area with a cloakroom beside the backdoor, accessing the garden and covered learning area; and a large floor to ceiling paper-mâché tree and mountain, complete with mountain goat, in the middle of the room.

Learning occurred in a variety of places quite naturally as the school used the outdoors frequently. There is a downstairs room for music or large art projects, a special clay area, and plans to create a woodworking area as well as a cooking facility. The outdoors are used extensively and seen as a legitimate learning environment that can be accessed at any time. As a teacher noted, "Learning is spread out quite a bit but about 30% outside or even more if you count lunch hours and the trip to school when I accompany students on their bikes." As stated earlier, the playground incorporated edible gardens, natural materials for building and playing with, composting and using rainwater, and a covered seating area so students can be outside in all weather. Further plans to develop the outside area as a natural learning environment with water and native plants and habitats for animals showed the active, ongoing development of the school.

The school days started in a very relaxed, open atmosphere with students taking the lead. When they came in, they chatted with other students, teachers, or volunteers or engaged themselves in activities that caught their interest. For example:

- **8:50 am:** Two students having a piece of pizza as they hadn't had time for breakfast yet; Dave conversing with Bob about tracks he saw with another student on their bicycle ride to school; Andrea came in and started drawing.
- **9:05 am:** Two students finish their snack and are jointly looking at a book; Bob is interested in tracks so he is encouraged to consult a reference book on the topic. A conversation on track identification develops so Bob is encouraged to go outside and find evidence; Andrea now drawing with a volunteer and discussing the habitat of the animals she was

drawing, incorporating these elements into the picture; another student beside her suggests they colour words they know blue, words they don't know green and numbers red; she carries on with her self-initiated task.
- **9:25 am:** Carol reminds the younger students of the workshop they did on money last week and suggests they might have fun setting up a shop. The girls who love drawing start drawing things they could sell, while others collect items and the girl who was working on words and numbers uses her activity to make labels and prices. The shop takes over and engages the children in interactive buying, selling, and making changes until the morning break.

Students and teachers didn't hesitate to go outside and stay focused on their work in all weather. Having appropriate clothing for this purpose seemed to be an accepted requirement.

Example Primary-Age Lesson

Teachers used a wide diversity of teaching methods encouraging the use of multiple intelligences. Subjects such as language arts, science, and math were integrated into a workshop by focusing on a topic. Active, hands-on learning was the basis of workshops and throughout a real love of learning was modelled and encouraged.

In an example workshop on trees, the teacher and students were continuing their building of a huge paper-mâché tree in the middle of the class. Encouraging self-directed learning, when the teacher asked students to write something about trees on a piece of paper and was asked how to spell a word or an answer, students were asked to think of where they could find that information. Students happily responded and found the answers themselves. The teaching was very transaction-based with a large degree of teacher/student negotiation and emergence in the learning. Many students went outside for inspiration, while others added pictures to their written descriptions, making and cutting out things to add to the tree. She was very open and accepting of divergent ideas on learning, allowing students to develop their own ideas on what they added to the tree.

By guiding and facilitating rather than teacher-directed transmission of information, when students lost interest, she was able to refocus and re-engage them through encouragement and a variation of activities. After the initial written activity, she invited students into a cosy corner for a guided imagery activity that explored the tree's interdependencies. One student declined and happily chose to read nearby. After the imagery, students were very excited to discuss their experiences. Interest focused on baby bird nests and what they could use for cushioning. Many returned outside to find materials, while others thought of things they could bring in from home. Projects were now underway for making nest cushions. The teacher started to orally tell a story, *Harry and the Roses,* about a bird unravelling a sweater to line its nest, when the student who had been reading excitedly ran to the library and found the book to read. The workshop culminated with students breaking up, often in pairs, to work on finding tree information and recording it in their tree books. This workshop was very ecologically oriented. Throughout the workshop, students were empowered to learn for themselves, choose from various activities and develop critical thinking skills. Also, the teaching materials were all either recycled or made from sustainable materials.

Evaluation and Assessment

The bioregional school considered assessment to be an evaluative, individual teaching tool rather than a means of grading and grouping students. The School Manual explains,

> In our un-graded programs, we continuously assess what children are newly achieving. They are not compared to other students, nor to some predetermined idea of what they should know by a certain age. Our goal is for students to progress along their unique learning paths at a rate that is right for them.

When teachers were asked how the curriculum was evaluated, they replied there was no formal testing. For both the provincial and bioregional curricula, they got feedback from students through observations and discussions, entering these on various checklists.

Progress wheels are used as a tool for recording student progress as individual, community, and bioregional/global learners. There is an Independent Skills Wheel that incorporates reading, writing, mathematics, and problem-solving skills, as well as learning skills such as transferring experiences and information to other situations; challenging oneself with confidence; investigating many options; and respecting and taking ownership of work.

The wheel for progress as a community leader focuses on respect for self and others. More specifically it incorporates the ability to: respect opinions of others; contribute to and share responsibilities in groups; accept and give advice; solve social problems with those involved; understand the consensus process and participates effectively; respect and develop own and other's ability to learn; ask specific and challenging questions of others and show appreciation of someone else's point of view.

The wheel for progress as a bioregional and global learner identifies "respects our community outside the school" and "enjoys learning about nature" as the central themes. These themes incorporate learning about a variety of ecosystems and species that inhabit them; ecological concepts of *cycling*, change, carrying capacity, energy flow, *interdependence*, *diversity*, and *community*; cultural diversity; natural and cultural heritage; and issues such as climate change and ecological sustainability, population, and development.

These wheels are consistent with their holistic, child-directed approach to learning, as the assessment is not developed according to separate subject disciplines. In this way the assessment is reflective of ecological metaphors as it focuses on student development, *emergence*, and an *integrated* view of learning rather than discipline-focused content. The themes also emphasize the bioregional and *community* focus of the curriculum, ensuring assessment is incorporated into all aspects of the curriculum.

Checklists are used to record individualized learning outcomes for each student as they are attained,

> Based on this record, teachers know in which areas individual students need encouragement and experience. Teachers can then provide individual and group activities to suit. Older students also use this information for their own planning.
>
> (School Manual, p. 8)

Other student evaluations are broken into personal as well as discipline-centred areas of working with others, individual and project work, language arts, mathematics, science and cultural studies, physical activities, music, and art.

Evaluations occurred in various ways. A teaching assistant identified discussions with other teachers and students as well as observations as the main forms of evaluation. Formal tests for special needs were used to help identify needs as necessary. In response to questions about how student evaluations occurred a parent replied, "The children's learning is evaluated through the use of report cards and charts to get a picture of where they are at as a whole." When asked if parents were involved in evaluations, she replied,

> We used to have meetings with kids to review and set goals and I sat in on those. They don't happen now as it was hard to get the time and parent involvement. Students now do their own evaluations to get a picture of who they are, what they want to learn and how.

The Headteacher noted that parent-teacher feedback tends to happen informally on a daily basis. Report cards are done co-operatively with goal setting. These goals affect what is taught.

Field observations verified that evaluation and assessment was obviously an integral part of the teaching and learning process. Teachers were actively engaged throughout the day in discussions with students about personal learning goals and achievements; with parents informally at the end of the day; or more formally with other teachers and volunteers to discuss individual students or various learning activities. All were obviously dedicated to this process as it was given so much time and consideration.

In line with their commitment to life-long learning and sense of emergence, the staff and directors felt there was a need to evaluate their overall progress. They wanted to know how they were doing in their bioregional education in promoting ecological literacy; and what steps they could take to improve. This highlights how they used feedback from students, parents, and colleagues to enable *adaptation* and *emergence* in the development of effective teaching and learning practices. As such, the bioregional teaching and learning approaches are consistent with and exemplify an eco-centric teaching philosophy.

Conclusion

The foundation of teaching and learning effectively in developing a sustainable society, where all thrive through healthy and effective socio-ecological relationships, is in replacing the mechanistic root metaphors that guide our teaching and learning practices with the ecological principles of sustainable living systems that learn and respond as conditions around them change. Einstein is paraphrased by Ram Dass to have said that the significant problems we face cannot be solved at the same level of thinking we were at when we created them (Conifold, 2018). The mechanistic mindset and the hidden root metaphors of the dominant mechanistic paradigm are imbedded in teaching and teacher education to the extent they are undermining our ability to effectively transform our educational practices. In bringing this to light, it is clear we need transformational paradigmatic change in teacher education and K–12 pedagogy and learning so as to support the emergence of empowerment, innovation, and creativity in teaching and learning in bringing sustainable education to life.

Many of these teaching and learning approaches described throughout this chapter are not new to teachers, but they are not able to be commonly practised. We instinctively know they make for good education. Many of these teaching and learning innovations have been tried and still exist here and there, but typically they are subverted and don't last because they have been naively grafted onto an unsupportive mechanistic framework. When they align with and are supported by the ecological principles of sustainable living systems as they are in The Green School and bioregional school, they give new inspiration, helping us realize they can become the norm when supported by an eco-centric, sustainable educational system, and continuous efforts to innovate in teaching training, teaching, and learning. This living-system educational framework built on systems thinking and sustainability principles supports and encourages what we know to be good education.

Rather than have pedagogical innovations be initiated and then subverted by our industrial, mechanistic mindset, an eco-centric educational paradigm will provide the necessary foundation and support to incorporate diverse worldviews; develop true partnerships in learning; create place-responsive, experiential learning in appropriate contexts enabling Nature to be our model, mentor and measure; foster systems thinking and community-based initiatives; and develop education based on creativity and innovation through locally relevant *adaptation* and *emergence*. The exciting thing is educators already know this works. As it's very difficult to consistently implement transformative teaching in the mechanistic system that is designed for and encourages transmission of information, and limited teaching approaches, it's time to provide educators with the competencies as well as a system that supports a diversity of good teaching and learning approaches.

The ecological principles of *the whole is greater than the sum of its parts, interdependence, community, diversity, cycling* and *feedback*, and *adaptation* and *emergence* are foundations of how all sustainable living systems function. Our human systems need to do the same if we are to become sustainable. When these principles are used as our mindset and framework for teaching and educational reform, we will not only address the mechanistic obstacles that have stood in our way in responding effectively to the institutional and socio-ecological issues we need to address, but we will also have an educational system that empowers teachers and students to be creative, innovative, and able to adapt to and develop solutions as we grow, learn, and conditions around us continue to change. Teaching and learning based on ecological principles help us reconcile our problematic colonial, cultural, and socio-ecological relationships imbedded in mechanistic constructs in moving forward to reconnect, rethink, reconcile, reimage, and redesign in re-story-ating our ways of being. In short, an eco-centric educational paradigm based on ecological principles creates a learning environment for students, student teachers, teachers, and society to thrive and becomes a learning society able to respond effectively to the challenges we face in developing sustainably.

Teaching in systems will help all students rise to their potential and equip generations of children to tackle some of the greatest challenges humanity has ever faced. By connecting to community, the natural world, and the systems that shape our global society, schools can foster hope in the future and empower students to find their role in shaping it. How we educate influences how we think. To create the schools of the future we can learn from the lessons of sustainable living systems. Sustainability is a learning process. To become part of the change, schools and teachers need to learn and evolve. To help start this transition, as we are all part of this interconnected system, all of us can leverage change, although change may not be easy. But if we develop a realistic, inspiring vision of what the future of school could be, and then invest in it, change will come. Investing in the future of school will be a journey of discovery. We have imagination and creativity, and the wisdom of living systems to guide us. It's time to invest in transitioning to sustainable education in the way we design and manage our educational systems, and how we teach and learn.

References

Bonnett, M. (2002). Education for sustainability as a frame of mind. *Environmental Education Research*, *8*(1), 9–20.

Bowers, C. (1995). *Educating for an ecologically sustainable culture*. State of New York Press.

Chawla, L., & Escalante, M. (2007). *Student gains from place-based education*. Children, Youth and Environments Center for Research and Design. https://promiseofplace.org/sites/default/files/2018-06/student%20gains%20fr%20PBE%20fact%20sheet%202%20Nov%202007%20web.pdf

Conifold (October 3, 2018). History of Science and Math Stack Exchange. (n.d.). *Did Einstein say "We cannot solve our problems with the same thinking we used to create them"*? [Online forum post]. StackExchange. https://hsm.stackexchange.com/questions/7751/did-einstein-say-we-cannot-solve-our-problems-with-the-same-thinking-we-used-to

First Nations Educational Steering Committee (FNESC). (n.d.). *First peoples principles of learning*. Retrieved from https://www.fnesc.ca/wp/wp-content/uploads/2020/09/FNESC-Learning-First-Peoples-poster-11x17-hi-res-v2.pdf

Foster, J. (2001). Education as sustainability. *Environmental Education Research*, *7*(2), 154–165.

Goodlad, J. (1984). *A place called school: Prospects for the future*. McGraw-Hill Book Company.

Gough, S. (2002). Increasing the value of the environment: A 'real options' metaphor for learning. *Environmental Education Research*, *8*(1), 62–72.

Gough, A., & Whitehouse, H. (2003). The "Nature" of environmental education research from a feminist poststructuralist viewpoint. *Canadian Journal of Environmental Education*, *8*, 31–43.

Holling, C. S., Gunderson, L. H., & Peterson, G. (2002). Chapter 2: Resilience and adaptive cycles. In L. H. Gunderson & C. S. Holling (Eds.), *Panarchy: Understanding transformations in human and natural systems*. Island Press.

Hopkins, C. (2012). Reflections on 20+ years of ESD. *Journal of Education for Sustainable Development*, *6*(1), 21–35. https://journals.sagepub.com/doi/abs/10.1177/097340821100600108

Howard, P. (2020). *Living schools and 21ˢᵗ century education: Connecting what and how with why* (C. O'Brien & P. Howard, Eds.). Living Schools: Transforming Education. Retrieved from https://mspace.lib.umanitoba.ca/server/api/core/bitstreams/fe48ad35-c49d-4ab7-941c-59d17582f355/content

Institute for Experiential Learning. (2020). *What is experiential learning?* https://experientiallearninginstitute.org/resources/what-is-experiential-learning/

Ireland, L. (2007). *Educating for the 21st century: Advancing an ecologically sustainable society* [Doctoral dissertation], University of Stirling]. STORRE. http://hdl.handle.net/1893/240

Kohn, A. (2002). How Not to Get into College: The preoccupation with preparation. Independent School. Retrieved from https://www.alfiekohn.org/article/get-college/

Kolb, D. A. (1984). *Experiential learning: Experience as the source of learning and development* (Vol. 1). Prentice-Hall.

Meadows, D. H. (2008). *Thinking in systems: A primer* (D. Wright, Ed.). Chelsea Green Publishing Company.

Naess, A. (1989). *Ecology, community, and lifestyle: An outline of an ecosophy.* Cambridge University Press.

O'Brien, C., & Howard, P. (Eds.). (2020). *Living schools: Transforming education.* Retrieved from https://mspace.lib.umanitoba.ca/server/api/core/bitstreams/fe48ad35-c49d-4ab7-941c-59d17582f355/content

Orr, D. W. (1994). *Earth in mind: On education, environment, and the human prospect.* Island Press.

Perry, M. (2023). *Sustainable futures in Africa.* http://cradall.org/content/unesco-uil-and-sfg-network-present-%E2%80%98pluriversal-literacies-sustainable-and-equitable

Rauch, F. (2002). The potential of education for sustainable development for reform in schools. *Environmental Education Research, 8*(1), 43–51.

Robinson, K. (2015). *Creative schools: The grassroots revolution that is transforming education.* Penguin Books.

Roth, C. E. (1992). *Environmental literacy: Its roots, evolution and directions in the 1990s.* ERIC Clearinghouse for Science, Mathematics, and Environmental Education. https://files.eric.ed.gov/fulltext/ED348235.pdf

Sauvé, L. (2005). Currents in environmental education: Mapping a complex and evolving pedagogical field. *Canadian Journal of Environmental Education, 10,* 11–37.

Senge, P. M., Cambron-McCabe, N., Lucas, T., Smith, B., Dutton, J., & Kleiner, A. (2012). *Schools that learn: A fifth discipline fieldbook for educators, parents and everyone who cares about education* (2nd ed.). Crown Business.

Sessions, G. (Ed.). (1995). *Deep ecology for the 21st century.* Shambhala.

Sterling, S. (2001). *Sustainable education: Re-visioning learning and change.* Green Books Ltd.

Sterling, S. (2010). Learning for resilience, or the resilient learner? Towards a necessary reconciliation in a paradigm of sustainable education. *Environmental Education Research, 16*(5–6), 511–528. https://doi.org/10.1080/13504622.2010.505427

Suzuki, D., & Knudtson, P. (1993). *Wisdom of the elders.* Bantam Books.

Uhrmacher, P. B. (1995). Uncommon schooling: A historical look at Rudolf Steiner, anthroposophy, and Waldorf education. *Curriculum Inquiry, 25*(4), 381–406. https://www.jstor.org/stable/1180016?origin=crossref&seq=1

United Nations Economic Commission for Europe (UNECE). (2012). *Learning for the future: Competencies in education for sustainable development.* Retrieved from: https://unece.org/DAM/env/esd/ESD_Publications/Competences_Publication.pdf

Part IV

Next Steps

Adaptation and Emergence in Transitioning to Sustainable Education

Whatever you can do or dream you can, begin it. Boldness has genius, power, and magic in it. Begin it now.

(Goethe)

DOI: 10.4324/9781003389590-11

8 Transitioning Organizational Structure, Administration, and Leadership

Transitioning to sustainable education takes vision and supportive, transitional leadership. Administrators at all levels in the system need to be engaged in professional development to develop an understanding of sustainable education and the leadership skills needed to collaborate with all multi-stakeholders in establishing the vision, the "why" of educating for sustainability. This leads to deeper understanding, commitment, and motivation for all to engage in the transformational process and develop a collaborative mission statement that will direct what needs to change and how, across and between all levels in the system.

Facilitating Change Through the Ecological Principles

Westley (2002) recognized that effective systems change can be facilitated by managing through ecological principles as experiments to learn from, within the organization at each level, out to the community, and up to the scales above in the nested panarchy. As Westley et al. (2006) highlight in their book, *Getting to Maybe: How the World is Changed*, focusing on the ecological principles of complex adaptive systems can guide this transition.

Interdependence

Thinking systemically helps all stakeholders see the bigger picture of how sustainable education will help individuals, communities, and societies develop sustainably, and it also helps everyone realize how integral they all are to our educational system transforming, whether they are students, parents, teachers, administrators, support staff, maintenance personnel, or community members. In developing a learning organization that can transition, administrators need to establish and support strong networks of interdependent relationships through effective administration and leadership across and between levels in the educational panarchy: between students, teachers, parents, the school, school board, maintenance departments,

DOI: 10.4324/9781003389590-12

employee unions, community, and Ministry of Education as well as across levels: student to student; teacher to teacher; school to school; district to district, as well as collaboration with Ministries of Education across provinces and nations. Building a culture of trust is central to this relationship building. Often the layers of trust exist more strongly at the school level where administrators, teachers, students, and parents work more closely. In developing a panarchical learning organization that fosters interdependent collaboration, building trusting relationships will help break down the industrial practice of maintaining silos and defence of boundaries, thereby encouraging a flow of information between scales in the panarchy.

Meadows (1999) recognized this flow of information as a very powerful leverage of change. To create this flow of information within this interdependent structure of relationships, and help break down barriers, those who are traditionally lower in a hierarchical system, such as students, parents, teachers, and to some degree principals, need to work in a safe collaborative culture to seek out and engage with those in power, making themselves a serious player and worthy ally (Westley et al., 2006). Equally, those at higher levels need to be open to collaboration, releasing their traditional top-down control. Counterproductive, often subtle disciplinary practices for stepping out innovatively or communicating with someone you are not allowed to be speaking with in the traditional top-down hierarchy, alters and impedes information flows. In transitioning from the top-down dominator model, recognizing and addressing subconscious levels of fear, control, and spirals of silence, we need to work with all stakeholders to bringing to light how to work with and change these typically hidden effects of the industrial schooling system. Once these barriers are illuminated and dismantled, information can flow effectively throughout the system.

Building positive social and socio-ecological relationships is critical in shaping behaviour and creating flow. In order for this to happen, recognizing and mapping interdependent relationships can be a good starting point and then inviting people to participate and develop ideas to strengthen these relationships. Administrators play a central role in supporting these relationships to enable multi-stakeholders to self-organize in the system based on interdependence and innovations at various scales. This extends to Principals supporting teachers and students to also take on leadership roles. Rather than relying on a central controller, the responsibility is distributed among and across all levels in the panarchy where multi-stakeholders can cluster together to share information and develop collective interest (Westley et al., 2006). Another important way administrators can help initiate and support teachers and students taking on leadership roles through self-organization is by opening up flexible scheduling options in the traditional timetable. Instead of scheduling subjects in separate timetable slots, whether that is for specialists at elementary or secondary levels, a timetable can be structured around interdisciplinary learning blocks. Within these blocks, students and teachers are free to develop interdisciplinary topic-based units, with far more time to collaborate and learn in relevant contexts beyond the classroom walls.

Community

Developing these interdependent relationships can be strengthened when emphasizing the ecological principle of *community* in schools, districts, and their wider local human and more-than-human communities. Community involvement at the national and regional scales, as well as local scales, will help ensure sustainable education is not being developed in a silo, but responding to societal and community needs and interests. This is where communication is key to inviting people in to be part of the conversations in developing the educational vision, and in creating diversified mission statements that are contextually relevant to the communities schools are situated in. School districts and school boards need to be given freedom to extend the national and regional sustainable education vision statements to include locally appropriate vision and mission statements through the engagement of local communities and stakeholders in vital discussions, identifying and monitoring desired goals and processes that resonate with local communities, while ensuring the culture and climate of their schools embody the principles of sustainability and well-being for all and 21st-century learning (Kay, 2020a). To facilitate this, administrators need to establish open and reciprocal flows of information with ongoing support over time.

Maintenance departments and support staff are also key contributors to this educational community. Often dealing directly with the interface between schools and the environment, through grounds, buildings, resources, energy, water, and waste systems, the maintenance departments and support staff work with the "hidden" curriculum. These aspects of the hidden curriculum need to become visible in consciously bringing the more-than-human world into discussions in recognizing our interdependence, developing more effective socio-ecological relations and situating education in communities beyond the school grounds and built environment.

Diversity

Being open to diversity helps administrators and leaders at all levels seek out differences of opinions and diverse worldviews so that pluriversalism is supported. Developing workshops to facilitate extensive and intensive exchanges will create opportunities to introduce pluriversalism that offers diverse perspectives so as to develop new and diverse ways forward. It is also important to recognize, respond to, and incorporate diverse issues from various perspectives and levels in the panarchy in transitioning away from a mechanistic educational system. To enable diverse perspectives to come forward, administrators need to develop clear reciprocal communication to illuminate obstacles, needs, and innovative ideas. These conversations can lead to setting diverse, collaborative targets and action plans for each level in the panarchy, working in diverse ways to administer and manage transitional innovations that address identified obstacles and needs.

Feedback

Transitioning to administration and leadership in a complex adaptive system depends on continuous *feedback* through effective reciprocal communication networks so everyone is aware of and can learn from innovations and changes at various levels. This is particularly important in a panarchy as changes at any level influence levels above and below. Meadows (1999) identifies driving positive feedback loops and regulating negative feedback loops as two of the top ten leverages of power to drive change in systems. As such, administrators need to incorporate and evaluate the effectiveness of various interdependent relationships with the community, within and across scales in the panarchy, and how well diversity is being incorporated. Information will be needed on how system processes such as communication networks within and between all levels in the panarchy enable effective feedback so as to monitor and improve both processes and outcomes. Westley et al. (2006) emphasized that to facilitate this process administrators need to set multiple information targets, not just performance targets that tell you to pause and look again. Asking who is involved, who is absent and why, what each player is learning about their own workplace, what changes each player has made, what lessons are players learning, and how they are sharing them will give administrators feedback and insights into how to adapt further. Teachers, Teachers On Call, parents, and learners play a key role in this feedback as they directly experience the impacts of transitional initiatives and are integral to *adapting* and supporting effective *emergent* changes.

This continual assessment of successes, obstacles, and needs based on multi-stakeholder engagement and *feedback* will help administration and leadership see what needs to change. Looking for and sharing what you see and hear can make sense of the patterns that are *emerging* and help maintain a systems perspective to develop effective next steps (Westley et al., 2006). Gathering feedback through an iterative process of engaging diverse multi-stakeholder perspectives, across and between scales, will help create both broad-based goals and umbrella strategies nationally and regionally, more specific challenges and strategies for local districts and schools, as well as professional development opportunities and the development of personal growth plans for administrators, teachers, and students. As these transitional initiatives, these creative strategies unfold they need continuous *feedback* to enable a continuous active learning cycle: from continually reflecting on the results of initiating change, making reflection the centrepiece in your actions, gaining feedback from others, to then responding to that feedback through further transitional iterations and feedback.

As the objective is to develop a collaborative community in transitioning to an adaptive living system, transparency and sharing feedback are essential. In this way, a systems perspective is supported. All will be aware of successes, obstacles, and needs; feel part of the transformative process; and will have insightful perspectives on next step as part of the greater whole, and be able to celebrating successes along the way.

Adaptation and Emergence

Developing effective leadership is needed to help all *adapt* and *emerge* innovative practices in transitioning to sustainable education practices. People tend to want order and answers, and change can be uncomfortable; so to help people become comfortable with *emergence*, effective, supportive leadership needed in creating a "safe-to-fail" culture, everyone needs to understand it is an evolving process supporting experimentation and creativity.

Once administrators, teachers, parents, and students develop an understanding of complex adaptive systems thinking, leaders can use Appreciative Inquiry to help support fellow administrators, teachers, student teachers, and students be change agents as it is a positive methodology of change based on individual and systemic strengths, successes, and resources. Through Appreciative Inquiry participants *define* the topic or area they wish to focus on; *discover* strengths, skills, and resources needed to foster change; *dream* or imagine an inspiring vision of their future; *design* concrete steps and actions; and finally realize their *destiny* by putting their strategies into action, and *revise* as necessary to make their vision a reality (Miglianico et al., 2023).

In developing a learning organization, some things will work, others may not and that's as it should be as we learn through trial and error, by experimenting how to *adapt* so as to *emerge* effective practices. In light of this, administrators need to provide space for the active learning cycle of direct experience, critical reflection, abstract conceptualization, and active experimentation as this creates a "safe-to-fail" culture of learning that all can thrive (Westley et al., 2006).

As innovations can influence change from any level in a panarchy, Biggs et al. (2015, pp. 269–271) make the following practical suggestions for transitioning management practice that will enable applications of these ecological principles:

- Clarify goals and develop and monitor relevant metrics for each principle.
- Take an integrative approach that builds on multiple knowledge sources.

- Shift away from exclusively managing for efficiency towards planning for uncertainty and surprise.
- Create spaces for spontaneous exploration.
- Build trust and social capital.

Applying these suggestions in developing a transformative learning culture will help administrators initiate and encourage change that leads to the whole becomes greater than the sum of its parts.

The Whole Is Greater Than the Sum of Its Parts

Taken together, transitioning the organizational structure, administration, and leadership of the educational system based on these ecological principles of *interdependence, community, diversity, feedback, adaptation,* and *emergence* leads to the most powerful leverage to initiate system change: changing the paradigm underlying education (Meadows, 1999). Throughout these chapters we have explored the importance of developing an eco-centric paradigm through transitioning from an industrial, mechanistic educational system to sustainable education. As noted above, developing a depth of understanding in an eco-centric paradigm, understanding sustainable education and panarchical systems thinking, where adaptive cycles of innovation at various levels can influence the emergence of further adaptations at levels above and below, are necessary professional development for leaders and administrators at all levels in the system in changing the paradigm. This helps everyone understand the larger context and enables the whole to be greater than the sum of its parts.

Having developed a clear vision and eco-centric understanding, administration and leadership need to establish the goal of the system as a learning organization to develop sustainable education to flourish. Establishing this goal is the second most powerful leverage of change (Meadows, 1999) as it develops a culture of learning to open opportunities for change across the system (Senge et al., 2012). Rather than controlling and managing people or other parts of the system to a predetermined expectation, it's important to manage patterns to create the climate within which a wide variety of strategies can grow (Westley et al., 2006). Creating the climate for a learning organization to enable this transformation is an essential first step. To do this, administrators set the conditions so learning can thrive. This helps everyone collaborate and find flow in developing an organic, supportive system that can adapt and emerge as conditions change and understanding develops.

As the distribution of power over the rules of the system is the third most powerful leverage of change (Meadows, 1999), developing adaptive co-management to facilitate transitioning to sustainable education allows administrators, from Ministries of Education to school district

Superintendents to school boards, and school Principals, to transition from a traditional top-down management style based on centralized control to one that enables collaboration between all stakeholders at all levels in the educational panarchy. This will help administrators create true learning organizations that are open to new ideas. To do so necessitates addressing leverages of power: linking discussions of power and redistribution directly to the organization's mission so that each level in the hierarchy has leverage for change (Westley et al., 2006). Support transformations that are initiated at other levels in the panarchy enables community members, teachers, parents, and students to all take on leadership roles.

Leadership

Throughout this transition to sustainable education, it is important to develop positive, effective leadership that can motivate and support all transitioning to sustainable education whether transitional ideas are initiated by administrators, community members, maintenance personnel, support staff, parents, teachers, or students. In recognizing the need for significant changes in education in the 21st century, a variety of leadership approaches have been seen to be effective (Kutsyuruba et al., 2023). Positive leadership, servant leadership, authentic leadership, and collaborative leadership are particularly effective approaches for administrators and leaders, at all levels in the educational panarchy, to transition from a traditional top-down, plan-and-impose model of leadership to one that empowers others and supports innovation across the system. Used in tandem, these leadership approaches model and enable the application of the ecological principles in transitioning administration for sustainability.

To begin with, servant leadership is an excellent concept and approach to help reposition administrators in transitioning their leadership approach away from planning and imposing change by supporting and mentoring innovations in teacher training as well as innovations generated by multi-stakeholders at all levels in the educational panarchy from classrooms to schools, districts, and regions. This approach helps develop humility and openness to collaboration with others.

In line with the criteria for transitioning administration and leadership through a shared vision, supported by the ecological principles of *interdependence, community, diversity,* and *feedback*, positive leadership facilitates *adaptation* and *emergence* by developing the sense of purpose, creating and building quality relationships, and the realization and utilization of talents, strengths, and gifts (Cherkowski et al., 2020). Roache et al. (2023, p. 113) note, "It includes promoting outcomes that will cause all educational stakeholders to thrive, flourish, and create positive energy at work, aiding in the sustaining of learning communities." Positive leadership, with roots in positive psychology, fosters believing in everyone's adaptive capacity and ability to innovate. In looking for and sharing extraordinary levels of achievement, and focusing on strengths and capacities to flourish and overcome obstacles, leaders work collaboratively to improve education, energize networks, and create the conditions for all to flourish. This positive approach recognizes leadership will motivate and empower others if it is based on gratitude, respect, compassion, and well-being (Esper, 2023). Other strategies to exemplify positive leadership include weekly internal memos that include positive encouragement for leadership initiatives at all levels in the panarchy, and recognition of successes being achieved, as well as keeping all informed of upcoming collaborative meetings and celebratory events.

Authentic leadership is based on inclusivity and respect, as well as collaborative and transparent decision-making so is instrumental in fostering *interdependence, community,* and *diversity.* Hetherington et al. (2023) advocate authentic leadership as they recognized through their research with schools that when multi-stakeholder voices are authentically involved in decision-making, it creates a welcoming atmosphere for collaboration and gives everyone ownership over the vision and its implementation.

The fourth pillar of effective leadership, collaborative leadership, recognizes the importance of working across networks of people within the organization and with other groups beyond the organization, building communities engaged in mutual learning and sharing, to address a complex problem (Squires & London, 2023). As such, collaborative leadership can be seen as systems leadership, a very important compliment to these other leadership approaches. Key to collaborative leadership is management and leadership providing a safe space for collaboration, and adaptive co-management, shifting away from expectations based on individualism. In support of this, Squires and London (2023) note that positive and collaborative leadership

> are described as transformational, whereby leaders engage the group or organization in generative dialogue and collaborative work to address complex issues and, thus, create positive change. Everyone is focused on a common purpose and committed to achieving the agreed-upon goals. Leaders establish safe spaces, underpinned by a sense of trust, creating space for generating potential solutions. The collective efficacy of the group is enhanced by the generative energy of members; the end result of the collective is superior to the amalgamation of individual efforts...Both leadership models can be scaled to the size of the organization. In essence, both types of leadership can be conceived of as systems leadership models where leaders work with organizational members horizontally and vertically throughout different hierarchical levels. (p. 268)

As we've seen, transitioning organizational structure and administration so that it can facilitate transitions in administration, curriculum, buildings, grounds and resources, teaching, and learning is a complex challenge, one that will depend on collaboration to empower multi-stakeholders across and between levels in the educational panarchy through positive, authentic, collaborative, servant leadership.

Managing Resources

One of the best ways to ensure success is to identify a reform strategy, spend the necessary organizational time and resources to roll out the reform initiative, and ensure the requisite time and financial resources to support the reform initiative are committed. Directly tying financial resources to desired outcomes tells people what is important and adds legitimacy to the organization's vision, mission, desired outcomes, and key initiatives (Kay, 2020b, p. 156). Rather than micro-allocating budgetary resources down to departmental or teacher levels Kay (2020b) recommends

> If instead, a principal was enabled to adopt a school-wide, holistic approach to budgeting – one that engaged teachers and students to prioritise school-wide initiatives and was directly linked to the school's financial resources to these initiatives, a significant increase in the effective use of financial resources would be realized. In my experience, the use of a holistic budgeting approach eliminates deficit and "use-it-or-lose-it" spending habits and fosters a school climate of efficiency and purpose that results in a surplus spending model. This, in turn, promotes creativity and innovation because students and teachers develop a sense of ownership, a clear understanding of how and why their financial resources are being utilized, and demonstrate greater guardianship over how money is spent. (p. 157)

This flexibility allows meaningful learning opportunities to emerge. Transportation costs for experiential place-based learning is often a limitation teachers come up against. A holistic approach to budgeting could help a principal support learning opportunities away from the school grounds, thereby enabling better access to the surrounding environments, improving interdependent relationships, *community*, and *diversity* of contextually relevant learning that emerges from inquiry-based learning.

Although this holistic approach to managing resources through budgeting is applied to the school level, it is equally important at higher levels in the educational panarchy. As initiatives and outcomes cannot be predetermined but emerge through decentralized pluriversalism, multi-stakeholder collaboration, and adapting as conditions change, regional and school district administrators also need freedom to adapt and allocate resources where needed in supporting initiatives as they emerge organically in transitioning to sustainable education.

Policy Development

As noted earlier, Donnella Meadows (1999), a foremost thinker in systems thinking and change, identified changing the goal of the system as the second most powerful leverage of change, behind changing paradigms. In support of the goal to transition to sustainable education based on an eco-centric paradigm, as a learning organization, policies will define and support changes, setting the conditions for all to thrive. For these transitions in administration and management to occur, administrators will need to work with government officials and other administrators at various levels in the panarchy to develop policies that encourage and support administrators transitioning their own practice, and supporting others to engage in the transitional process. As noted earlier, policies based on implementing sustainable education through ecological principles will necessarily transition from those based on *efficiency, control, constancy, and predictability*, hallmarks of traditional top-down management systems, to those that embrace *unpredictability, change, adaptiveness, and persistence* (Holling & Gunderson, 2002). This will enable administrators and leaders at all levels to facilitate and support transitions in complex adaptive systems within an evolving environment.

As noted in Chapter 4, policies are needed to support professional development, collaboration, and the resources to transition to this living systems approach to education based on the innovations highlighted by the ecological principles. For *interdependence* to be enhanced, policies that encourage and support collaboration will lead to developing stronger relationships and networks as innovations will, by policy design, need to be based on collaboration and cross-scale linkages. For this collaboration to occur, time and resources need to be built into policies as time is typically voiced as an obstacle to working with others. Such policies will support budgetary requirements for teachers to have release time from their classroom teaching, and administrators from their managerial tasks so they can be part of professional development, professional learning communities within schools and districts, as well as collaborative meetings at other levels in the panarchy, particularly in the early transitional stages and when new initiatives are planned.

These collaborative policies will also need to highlight the importance of *community* by identifying *diversity* as a requirement: representing diversity in gender, culture, as well as Indigenous worldviews to ensure pluriversalism in these *interdependent* relationships. This ensures *community* consultation and involvement.

As all buildings, grounds, and resources need to transition to sustainable socio-ecological relationships, administrators will also need to develop policies with the government funding agencies to provide both funds and resources; and with the maintenance department to develop initiatives to transition, energy, water, waste, and materials year-by-year. Policies are needed to use sustainable, renewable resources and develop the school grounds and create *community* access. As the more-than-human environment is integral to developing sustainable education in terms of where learning occurs, policies are also needed to be negotiated to enable learning beyond the classroom walls and school grounds. Teachers often face financial obstacles when transportation is needed for place-based experiential learning away from the school, limiting access and connection to natural environments. Overall, once there is a clear policy that all schools need to model and teach sustainability, changes will happen and be supported.

Conclusion

As we've seen throughout this chapter, transitioning organizational structure, administration, and enabling diverse supportive leadership at all levels in the panarchy can create the structure needed to transition to sustainable education. Applying the ecological principle of *interdependence* focuses administrative efforts on creating and maintaining effective, trusting relationships across and between scales. This enables powerful interactive *communities* to embrace the *diversity* needed to develop resilient learning communities that represent and draw on pluriversal perspectives. In response to local, regional, national, and international societal needs and interests, *feedback* is key to ensure information flows to support effective communication and dialogue, supporting *adaptation* and *emergence*. When all multi-stakeholders in the educational system have a voice and are active participants in this learning organization, innovation and creativity are encouraged and supported in *adapting* and *emerging* creative innovations and solutions through a "safe-to-fail" culture of regenerative, emergent growth. In this way, the organizational structure, administration, and leadership creates the conditions for *the whole to be greater than the sum of its parts*. This, in turn, will help the entire system in buildings and grounds curriculum, teaching, and learning transition so all will thrive through sustainable education.

References

Biggs, R., Schluter, M., & Schoon, M. (2015). *Principles for building resilience: Sustaining ecosystem services in social-ecological systems.* Cambridge University Press.
Cherkowski, S., Kutsyuruba, B., & Walker, K. (2020). Positive leadership: Animating purpose, presence, passion and play for flourishing in schools. *Journal*

of Educational Administration, 58(4), 401–415. https://doi.org/10.1108/JEA-04-2019-0076

del Carmen Esper, M. (2023). Building bridges for the whole child education and flourishing school community: A case study of effective principal leadership in Benjamin Kutsyuruba. In S. Cherkowski, & K. D. Walker (Eds.), *Leadership for flourishing in educational contexts*. Canadian Scholars.

Hetherington, R., Haley, C., & Spence, B. (2023). Wellness as foundation for change: Intersection of leadership and trust. In B. Kutsyuruba, S. Cherkowski, & K. D. Walker (Eds.), *Leadership for flourishing in educational contexts*. Canadian Scholars.

Holling, C. S., & Gunderson, L. H. (2002). Resilience and adaptive cycles. In L. H. Gunderson, & C. S. Holling (Eds.), *Panarchy: Understanding transformations in human and natural systems*. Island Press.

Kay, B. (2020a). Supporting living schools through transformative governance and leadership: A Vermont experience. In C. O'Brien & P. Howard (Eds.), *Living schools: Transforming education*. Retrieved from https://mspace.lib.umanitoba.ca/server/api/core/bitstreams/fe48ad35-c49d-4ab7-941c-59d17582f355/content

Kay, B. (2020b). Leadership: Creating a culture of leadership to support living schools. In C. O'Brien & P. Howard (Eds.), *Living schools: Transforming education*. Retrieved from https://mspace.lib.umanitoba.ca/server/api/core/bitstreams/fe48ad35-c49d-4ab7-941c-59d17582f355/content

Kutsyuruba, B., Cherkowski, S., & Walker, K. D. (Eds.). (2023). *Leadership for flourishing in educational contexts*. Canadian Scholars.

Meadows, D. (1999). *Leverage points, places to intervene in a system*. The Donella Meadows Project: Academy for Systems Change. https://donellameadows.org/archives/leverage-points-places-to-intervene-in-a-system/

Miglianico, M., Goyette, N., Dubreuil, P., & Huot, A. (2023). Appreciative inquiry in the classroom: A new model for teacher development. In B. Kutsyuruba, S. Cherkowski, & K. D. Walker (Eds.), *Leadership for flourishing in educational contexts*. Canadian Scholars.

Roache, D., Thomson, S. B., & Marshall, J. (2023). Positive leadership approaches: Principles and practices for flourishing schools. In B. Kutsyuruba, S. Cherkowski, & K. D. Walker (Eds.), *Leadership for flourishing in educational contexts*. Canadian Scholars.

Senge, P. M., Cambron-McCabe, N., Lucas, T., Smith, B., Dutton, J., & Kleiner, A. (2012). *Schools that learn: A fifth discipline fieldbook for educators, parents and everyone who cares about education* (2nd ed.). Crown Business.

Squires, V., & London, C. (2023). Collaborative leadership as an approach to promote well-being on post-secondary campuses. In B. Kutsyuruba, S. Cherkowski, & K. D. Walker (Eds.), *Leadership for flourishing in educational contexts*. Canadian Scholars.

Westley, F. (2002). The devil in the dynamics: Adaptive management on the front lines. In L. H. Gunderson & C. S. Holling (Eds.), *Panarchy: Understanding transformations in human and natural systems*. Island Press.

Westley, F., Zimmerman, B., & Patton, M. (2006). *Getting to maybe: How the world is changed*. Random House Canada.

9 Transitioning Buildings, Grounds, and Resources

In advancing sustainable education through the hidden and overt curriculum, we need to transition our buildings, grounds, and resources so they model and enable teaching sustainability. Procurement policies and budgets need to be updated to support purchasing non-toxic, biodegradable cleaning products and all furnishings need to be replaced as needed with sustainably made alternatives. Energy systems need to take advantage of renewable energy systems. Incentives and bulk purchases by governments can make this more viable for school districts.

This is happening in a number of places at all levels of education. A few examples are School District 10, Arrow Lakes in British Columbia (BC), Gulf Islands Secondary School, Salt Spring Island, BC as government-run elementary and secondary schools; Dawson College, Canada and Oberlin College in the United States for post-secondary education; and The Green School, Bali for K–12 education as an exemplary independent school.

Arrow Lakes School District, BC, Canada

Paying attention to transitioning buildings and resources was exemplified by the former head of the Maintenance Department in School District 10, Arrow Lakes in British Columbia, as he was personally committed to sustainability. In the early 2000s he made sure all cleaning supplies were sustainable and he replaced worn carpeting with Interface Flor carpet tiles, which are sustainably produced, mimic the forest floor, and can be installed as needed whenever older carpeting wore out, without replacing the whole carpet. This particular carpet company also takes back used carpeting so they can use the recycled fibres in new products. Energy for the school was also redesigned by pioneering a parallel biofuel energy system, partnering with the local logging industry to used waste wood. This system could augment and hopefully replace the oil-based heating system. Subsequently, students in the environmental club raised funds for a solar panel with the help of a well-established recycling programme, and through a whole-school

DOI: 10.4324/9781003389590-13

initiative established a school garden and tree planting on the grounds to help with drainage and a wind block on the playing fields. Through these initiatives from the Maintenance Department and collaboration between the students, teachers, administration, and school board, the school's hidden curriculum is reinforcing *interdependence* with the natural environment, partnership with the *community, diversity* of initiatives, *cycling, feedback, adaptation,* and *emergence.*

Gulf Islands Secondary School, BC, Canada

School District 64, on the Southern Gulf Islands in British Columbia, Canada is also a pioneering example of putting many sustainable initiatives into practice. The high school students and the local school district partnered with the Salt Spring Island Community Energy Group and took on the task to bring solar power to the local high school, with the school's environmental club involved in crafting the initial solar proposal to the school board. The 84 solar photovoltaic panels covering the Gulf Islands Secondary School gym roof were purchased with community grants and local business and citizen support, going live in December 2014. At the time it was the largest school-based solar array in British Columbia, and according to B.C. Hydro (n.d.), and the seventh largest overall. An innovative aspect of this project is the money the school saves on its energy bill from the 21-kilowatt project supports an annual $2,000 scholarship for a graduating student planning to study renewable energy or climate change. A government grant initiated funding, and in gaining community support for the project, volunteers went to businesses around town dressed in graduation garb with solar panels on their mortar board hats, collecting donations ranging from the purchase of a solar panel to any small amount an individual community member could afford to contribute. In this way, contributions ranged from thousands of dollars down to change people had in their pockets. In addition to community funding a renewable energy company, a few wealthier residents provided significant matching grants to raise the necessary funds. In this way everyone felt part of this community initiative. The learning opportunity goes beyond the scholarship, said School District 64 superintendent Lisa Halstead, "I think it's an absolutely terrific project. Not only does it provide energy and scholarships for our students, but we can also use it as a teaching tool in the district" (Times Columnist, n.d.).

This sustainable energy project is one of several projects at the school encouraging sustainability. The school has a greenhouse, as well as a "living lettuce wall" that grows produce for the cafeteria's salad bar. Working with the school's chef in the greenhouse is an elective programme and students sell fresh salad lunches, made from produce grown in their greenhouse and living lettuce wall, to students, staff, and the community from a mobile food truck outside the school. In 2023 the school district purchased electric school buses, further advancing their transition and commitment to model and teach sustainability. These initiatives show exemplary partnerships between the school and community, as well as between levels in the educational pan-archy from students, teachers, administrators, the community, and school board. When taken as a whole, these transitional developments, developed from the continuous efforts of many in overcoming various obstacles, exem-plify how the ecological principles of *interdependence, community, diversity,* cycling, *feedback, adaptation,* and *emergence* are keys to their success.

Dawson College, Quebec, Canada

At Dawson College they are repurposing education so that it contributes to "wellbeing for all – individually, collectively, and for the 'other than human' life on this planet" (O'Brien & Howard, 2020, para. 1). They have done a lot of work developing a living campus for reconnecting people, com-munity, and Nature by enhancing the campus for wildlife and creating with various gardens for people and pollinators so as to extend learning spaces beyond classroom walls. The Sustainability Co-ordinator works to connect students, the community, and instructors to help integrate sustainability and the campus into their courses. As Adam and Cassidy (2020) note,

> We define our College as a Living Campus, which embodies our jour-ney towards human and ecological well-being, and which includes human health and happiness, social justice, responsible economic activity, and a healthy natural environment for current and future generations. Defining ourselves as a Living Campus involves concep-tualizing the College as a learning platform that breaks down tradi-tional classroom structures and uses the entire campus as a learning laboratory. (p. 172)

These initiatives are powerful examples of aligning the hidden and overt cur-riculum through transitioning to a living campus creates a strong community in recognizing interdependence between our human and more-than-human species. *Diverse* initiatives are constantly being initiated by the Sustainabil-ity Co-ordinator and staff. Based on a collaborative *community, feedback* is constantly sought so they can continue to *adapt* and *emerge* new initiatives.

Oberlin College, Ohio, USA

At Oberlin College, David Orr launched the green campus movement or-ganizing students to study energy, water, and materials use on college cam-puses and in 1995 spearheaded the effort to design the first substantially green building on a US college campus. In developing the project, Orr (2021) summarized,

> [they held] thirteen design charrettes open to all students and to the wider community to build a knowledgeable constituency for the project and to calibrate the design to the larger purposes of the Environmental Studies Program. The project evolved into an experiment in applied hope and ecological design that included on-site water purification, solar technology, use of non-toxic materials, gardens, land restoration, and accurate feedback on building performance.

By making various aspects of sustainability visible, practical, and affordable, they exemplified what is possible. For example, The Adam Joseph Lewis Center purifies all its wastewater through a living water system where plants and fish filter and purify the water in successive stages, and it is the first college building in the US powered entirely by sunlight.

> The Adam Joseph Lewis Center is a milestone building. With ingenuity and effort, it provided a positive message that we can design neighborhoods, communities, cities, and nations to enhance the biosphere. It continues to provide many lessons for designers and students (para 1)
>
> (Orr, 2021)

The building was intended to be not just a place where classes were held, but a place that educated by its design, operation, and evolution. Architecture, in other words, should be thought a kind of crystalized pedagogy that instructs one way or another (Orr, 1994). Building and landscape design can reveal the larger ecological context in which it exists, can inspire hope, build practical competence, and help generate momentum towards a better and more coherent world (Orr, 2021).

Orr (2021) further elaborates,

> … the most important and perhaps the most difficult to measure result of the project was its effect on students, both those who participated in the design phase and the thousands who subsequently took classes in a solar-powered, zero-discharge, non-toxic ambience of innovation, risk-taking, and leadership. Many went on to various careers inspired by the architects, engineers, and designers who made hope for a better world tangible and inspiring. (para 10)

A pioneering advocate for sustainability education through his numerous publications, David Orr, through his work at Oberlin College, has initiated these powerful transitioning projects so we not only teach but model the ecological principles in practice.

The Green School, Bali

At The Green School, Bali, that caters to students from pre-Kindergarten to Grade 12, the typically hidden curriculum of buildings, grounds, and resources has become intentionally part of the overt curriculum. The buildings are made of sustainable local bamboo, beautifully designed to blend into the natural environment, incorporating natural lighting, fresh breezes, and open access to the outdoors with a wall-less campus. Understanding the importance of contextually relevant, place-based learning, the school empowers students to learn outside, in gardens that grow their own food, in the local communities, as well as in classrooms, in developing sustainability initiatives, such as micro-hydro energy from the local stream. Over the years the school has developed solar and hydro power with a vision to become carbon positive, closed loop waste management systems, including composting for their gardens, an aquaponics centre that integrates fish and growing plants, and a biobus to provide sustainable transportation for students and community members (Green School Bali, n.d.).

The Green School Bali recognizes how interdependent the buildings, grounds, community, and natural environment are with effective sustainable education. Through their intentional design of the buildings and campus, and extending learning to the surrounding community and more than human environment, they are exemplifying the ecological principles of *interdependence, community, diversity, feedback, adaptation,* and *emergence.*

Conclusion

Examples such as these are occurring in many school districts and schools in many countries. Numerous organizations such as Eco-Schools (n.d.) and Learning for Sustainable Futures (LSF) (n.d), to name a few, provide programmes, information, and sources for grants to help encourage students and teachers to green their schools and grounds in transitioning to sustainability. These are all excellent initiatives that apply the ecological principles to transition aspects of the hidden curriculum, particularly as they support transitions in curriculum, teaching, and learning.

References

Adam, C. & Cassidy, R. (2020) Nature and place-based orientation: Wellbeing for all—A story of Dawson College's Living Campus. In C. O'Brien, & P. Howard (Eds.), *Living schools: Transforming education.* ESWB Press.

Retrieved from https://mspace.lib.umanitoba.ca/server/api/core/bitstreams/fe48ad35-c49d-4ab7-941c-59d17582f355/content

B.C. Hydro (n.d.). *Salt Spring's 'solar high' gets power, and a scholarship, from sun*. https://saltspringsolar.ca/bc-hydro-salt-springs-solar-high-gets-power-scholarship-sun/

Eco Schools. (n.d.). https://www.ecoschools.global

Green School Bali. (n.d.). https://www.greenschool.org

Learning for a Sustainable Future (LSF). (n.d.). https://lsf-lst.ca

Orr, D. W. (1994). *Earth in mind: On education, environment, and the human prospect*. Island Press.

Orr, D. (2021). Building as pedagogy: Oberlin's Adam Joseph Lewis Center. *Buildings & Cities*. Retrieved from https://www.buildingsandcities.org/insights/research-pathways/building-pedagogy.html

O'Brien, C., & Howard, P. (Eds.). (2020). *Living schools: Transforming education*. Retrieved from https://mspace.lib.umanitoba.ca/server/api/core/bitstreams/fe48ad35-c49d-4ab7-941c-59d17582f355/content

Times Columnist. (n.d.). *Salt Spring Island school warms up to solar*. http://www.timescolonist.com/news/local/salt-spring-island-school-warms-up-to-solar-1.1820869

10 Transitioning Curriculum

Inspiration from the Ecological Principles

In efforts to transition the mechanistic provincial curriculum to one based on sustainability, teachers can use the ecological principles to inspire decentralized curriculum development, ideally supported by policy and administration at higher levels in the panarchy. Initially, until such policies and support are in place, inspired teachers may develop a unit of study independently, ideally with the help of the students, by integrating traditional subjects into a theme or topic with the help of the Eco-Centric Curriculum Framework as described in detail in Chapter 6. By applying each of the ecological principles, these initial efforts can help transition our industrial curriculum to one that will support sustainable education and develop sustainability as a frame of mind.

Interdependence and Community

As educators work in a panarchy, where change can come from any level in the system, motivation to transition curriculum can come from a teacher, student, parent, community member, principal, superintendent, school board, or the Ministry of Education. Wherever the initial motivation comes from, innovations can cascade through the system to motivate others in the educational community.

As effective sustainable education curricula need to be contextually relevant and place-based, teachers will need to be involved in grassroots, decentralized curriculum development that connects to community and student needs. A teacher may initially start the transitioning process in their own classroom based on localized needs and co-creating with students; and then through sharing their experiences with their colleagues, teachers can inspire others to get involved in the transitioning process. Initiating change in one class can influence other teachers at the same school, as well as teachers at other schools across the higher district level in the panarchy, thereby creating opportunities

DOI: 10.4324/9781003389590-14

for a systemic cascade of changes. As project-based inquiry learning incorporates systems thinking, it often leads to student action projects that can influence supportive changes in administration and leadership, as well as buildings, grounds, and resources. For example, a project on energy can ultimately lead to students partnering with community organizations to gain support from the Principal, school board, and maintenance department to install solar panels on the school roof, and purchase electric school buses as they did on Salt Spring Island, School District #64, British Columbia, Canada.

Starting with just one unit can help teachers gain a little experience and see how motivated students become, particularly when they are involved in planning. The initial inquiry-based unit may be focused on a big idea or an area of content in science or social studies at the elementary level, or a particular subject at the secondary level. In developing a curriculum unit, the teacher and students should be looking for ways to develop learning outside with the surrounding community members, both human and more-than-human. Once the unit develops through the Eco-Centric Curriculum Framework, further interdisciplinary, diverse, and community-based learning opportunities arise.

If collaboration isn't initially possible, an individual teacher can still develop a unit that is open to incorporating learning opportunities that relate to other subjects, showing how interdependent their subject is in real-world contexts. Relating topics to community needs in transitioning to sustainability, and inviting input from the students as to what aspects of the topic they are particularly interested in, as well as how they would like to go about learning, empowers the community and students to be co-partners in developing relevant learning opportunities. Topics related to science or social studies can integrate English and other languages, mathematics, and fine and performing arts in expressing their learning; while topics starting from mathematics, language, and the fine or performing arts can use science or social studies as contexts for learning and expression. Consulting with colleagues can help integrate content a teacher may not typically teach, and pique the interest of others.

As teachers see the opportunities to weave in other subject disciplines, the unit-based learning can be expanded into half or full days as social studies, science, language arts, mathematics, and other subject disciplines offer diverse ways to extend and deepen the learning. As teachers become more experienced, open, and confident in topic-based experiential learning, teaching and learning transition to become more relevant and contextualized rather than fragmented and abstract. Dissolving the traditional discipline-centred timetable opens opportunities for integrated inquiry-based learning beyond the classroom, providing the time to learn authentically in context.

At secondary as well as elementary levels of education, the most advantageous ways of developing interdisciplinary curricula are to collaborate with other colleagues. To start with, two or three colleagues can get together to co-plan a unit that integrates their various subjects around a big idea or question of interest to help students develop various competencies and systemic knowledge. Creating a professional learning community with even more teachers across all disciplines will offer even greater opportunities to develop holistic curriculum units based on the Eco-Centric Curriculum Framework.

Teachers advocating for timetable changes to create interdisciplinary learning blocks is an important way to collaborate with administrators in supporting this transitional movement. Teachers and administrators likely need to also advocate with their union as well. As is often the case, it is the collective agreement that drives the timetable in that it typically incorporates language that limits how much preparation time a teacher is allotted, how many actual students they can have, the number of classes that can run outside the timetable, and what courses a teacher is allowed to teach based on qualifications or seniority. As an interdisciplinary curriculum involves teaching a variety of subjects or the collaboration of a number of teachers teaching across traditional subject boundaries, these collaborations, that enable systems thinking, and new ways of teaching and learning, are essential.

Diversity

As noted above, in trying out this curricular approach, it's advantageous for a diverse group of teachers to work together through a professional learning community to develop a sample unit, but diversity also relates to integrating diverse competencies and ways of learning. In developing learning opportunities teachers need to address:

- student developmental needs;
- multiple intelligences;
- critical thinking;
- reciprocal feedback
- enabling creativity and innovation;
- experiential learning in diverse human and more-than-human communities;
- systems thinking in being aware of how ideas and subjects are interdependent;
- diverse ways of expressing learning;
- integration of government curriculum objectives; and
- identifying how ecological principles can be integrated as the core of the learning as the unit develops.

Weaving in the active learning cycle detailed in Chapter 7 helps teachers integrate these diverse considerations and ground curriculum development in inquiry-based experiential learning. This curriculum design also allows for emergent learning that is not pre-planned and imposed from the top-down, from teacher to students. This makes the curriculum open to diversity rather than insisting on conformity.

Cross-referencing activities with prescribed competencies and/or learning outcomes in various subject disciplines can ensure the diversity in the mechanistic curriculum is being addressed and suggest ways to further extend learning to develop systems thinking. This will also help the teachers assess learning related to diverse competencies while developing curricula that is contextually relevant, helping students understand how various subject-related content relates to the greater whole.

Cycling, Feedback, Adaptation, and Emergence

As transitioning to decentralized curriculum development is a process, educators and administrators need to let go of *efficiency, control, constancy, and predictability* to embrace *unpredictability, change, adaptiveness, and persistence.* To enable this transition, teachers need to rely on and incorporate *feedback* from students, parents, colleagues, the maintenance department, and administrators. To help develop a learning perspective themselves, it benefits educators to constantly assess initial efforts for successes, obstacles, and needs so as to improve curriculum development as an iterative learning opportunity of *adaptation* and *emergence*. As learning is an emergent process, curriculum developers need to be open to unpredictable *feedback*, willing to *adapt* original plans, thereby avoiding the mechanistic approach of planning and imposing curriculum from the top-down. In being open, *adaptive*, and place-responsive, curriculum planning needs to leave space for uncertainty by embracing *adaptation* and persistence, so as to enable *emergence* – often in unexpected ways. Through this process of being open to *feedback, adaptation* and *emergence* "teachable moments" are integral rather than isolated opportunities.

The Whole Is Greater than the Sum of Its Parts

In transitioning curriculum, administrators and teachers need to see how *the whole is greater than the sum of its parts;* how the competencies and big ideas that come from topic-based teaching help students understand the greater whole, informed by the various subject disciplines. Each of the various subject disciplines are lenses or ways of seeing a topic such that each contribute to a more holistic understanding of a particular topic. For example, although climate change is often taught in science classes, social, geographic, mathematical/economic, and linguistic messaging and communication lenses are essential in understanding it, its impacts and implications, and how and why we are and are not dealing with it. Professional development and collaboration between teachers and administration, at various levels, are needed to understand how essential interdisciplinary learning is to progressing sustainable education, and enable innovative, decentralized curriculum development to bring it holistic learning to life for the students.

Integrating ecological principles brings out biomimicry as an important concept at all levels, from kindergarten to teacher training. Through biomimicry we learn to see ourselves as part of Nature, learning how to better adapt by seeing Nature as our model, mentor, and measure (Benyus, 2014). In this way, discipline-related learning emphasizes learning from Nature as well as interdisciplinary connections. As an example, social studies look at how social systems can mimic natural systems, informed by science, while science develops an understanding of natural systems and the connections between science, technology, society, and the environment.

Practical Examples

There are numerous examples of educators developing eco-centric curricula, connecting students to their local environments through interdisciplinary learning. Here are just a few examples representing various levels in education: a primary-level unit from a practising teacher and graduate from Royal Roads University's Master of Environmental Education and Communication program; a whole-school approach from case study research with a bioregional school; and a multi-level ecosystem approach from the UNEP-IEMP that is relevant for educators in diverse ecosystems across the world.

An Emergent Topic-Based Primary Unit Plan for Trees

By Alisha Underwood, Practising teacher and MA in Environmental Education and Communication, Royal Roads University

As a practicing primary teacher in British Columbia, Alisha Underwood was interested in transforming her mechanistic approach to teaching based on ecological principles. Her master's research at Royal Roads University centred on developing an in-depth unit of study she could start the academic year with, introducing the students to experiential, inquiry-based learning focused on the topic of Trees. Cycling through the Eco-Centric Curriculum Framework (Figure 6.1), she designed activities related to Trees and People, Trees in the Elements, A Tree's Life, and Trees and Animals Living Together. In each of these modules she identified how the ecological principles are incorporated through the activities, outlined specific activities to guide learning. Activities are intended to be taught in a holistic and emergent way, guided by the students' interests. These activities and inquiry questions will help students form an authentic connection with the local trees, guide them to develop a deep appreciation for trees, and expose the interdependence of the ecosystem. Curricular competencies from the BC kindergarten curriculum are naturally integrated throughout this unit and highlighted at the end of each activity.

At the end of the unit, she details how the teacher can assess learning as well as each ecological principle. Although this unit appears very teacher-focused, she recognizes emergent learning should occur and the teacher needs to stay open to follow student interests. As such, it is a helpful example of how teachers can initiate their teaching and learning transition to an eco-centric approach. This unit is detailed in the Appendix.

In teaching this unit, as well as others she developed throughout the year, and then sharing them with her colleagues, Ms. Underwood hopes to inspire her colleagues and administrator to join her in transitioning to sustainable education.

A Bioregional Approach

In attempting to work with the traditional provincial curriculum, the bioregional school in my case study research developed a hybrid model, blending a traditional government curriculum with a bioregional approach.

The School Manual and Handbook identified the curriculum as being made up of their own bioregional curriculum and the provincial government's curriculum guidelines and intended learning outcomes (ILOs). When directors, staff, students, parents, and volunteers were interviewed about the focus of the school curriculum, it became apparent that most had a dual conception of the school curriculum: the provincial curriculum and the bioregional curriculum, although all clearly felt the bioregional aspect was the "raison d'etre" and core of the school. The school founders referred to the entire school curriculum as bioregional education.

The bioregional curriculum is defined in terms of the four competency categories: Taking Care of Self, Building a Community, Knowledge of Our Bioregion, and Global Understanding. In Taking Care of Self the curriculum focuses on self-awareness, confidence, skills, knowledge, and understanding one's place in Nature. The category of Building a Community brings out participation, co-operation, meeting the needs of the community, understanding natural communities, and human reliance on the natural world. Knowledge of Our Bioregion incorporates components of the bioregion and ecological principles as well as knowledge of cultural, social, and economic patterns. The fourth level of Global Understanding brings out global components and systems as well as global issues and concerns.

Prioritizing the bioregional curriculum as the core, the provincial curriculum, based on separate subject disciplines, is integrated through teacher workshops and project-based learning. The centre of the curriculum diagram identifies "Children's Natural Curiosity and Innate Need to Learn" as the core. Around the core of child-directed learning are the following: "Through a holistic approach with learning occurring naturally and in context" and "Students need not be aware of traditional subject areas." Within the edges of the circle are independent learning skills such as communicate, inquire, understand, conceptualize, experiment, problem-solve, seek new perspectives, plan, and take action. On the outside of the circle are skills, knowledge, concepts, and understanding as well as the traditional subject areas of science, social studies, mathematics, reading, writing, physical education, drama, art, and music. This diagram is very telling in how deeply rooted a holistic, ecological view is in the bioregional curriculum. Holism, integrated critical thinking, and community are conceptual ecological metaphors that resonate throughout. The focus is on the student becoming a self-directed, *independent* learner in a *community* context. This precludes learning any particular subject matter.

By focusing on the child's natural curiosity and innate need to learn, and in being purposely child-directed, the intended and enacted curriculum is open and responsive to individual and community needs based on negotiation and consent. This is also the case with their emphasis on local, personal, applied and first-hand knowledge, and *interdisciplinary* learning. The Manual states,

> Learning is an integrated process. Problem solving, critical thinking and cooperation are life skills that transcend all subject areas. More importantly than specialized knowledge, students need to know how to learn and how to make connections amongst the things that they know. Integrated studies are well suited to developing a mature understanding of the world and an ability to think clearly to affect social change.

This explains and summarizes well the school's bioregional curricular emphasis, one that resonates with an ecological rather than mechanistic paradigm focused on discipline subjects.

The Provincial Curriculum

In order to receive 50% funding from the Provincial Ministry of Education, the school incorporates the provincial Learning Outcomes, outlined in the provincial Integrated Resource Packages (IRP) for each major subject area. As the government does not specify how these learning outcomes must be taught, or what texts must be used, the School Manual states, "Teachers use these documents and the learning outcomes in them to prepare individualized activities and to provide encouragement to students". The School

Manual describes how the B.C. Curriculum is perceived, interpreted, and intended to be incorporated into the overall school curriculum:

> [The] learning outcomes are met by students at their own pace. Topics in some subject areas are covered in a four-year cycle as well as when individuals show an interest. In this way all topics will be covered at least once every four years.

> Programs in Mathematics and English Language Arts are developed individually based on readiness, interest and prior understanding. Children are constantly learning – always constructing meaning, inventing and discovering skills, concepts and ideas. When a child is interested and ready, learning is quick, enjoyable and meaningful, not forced or frustrating.

While teaching the provincial curriculum, the school tries to avoid the emphasis on specific discipline learning outcomes at specific grade levels by focusing on individual needs and interests. However, the mechanistic subject-based learning outcomes presented some challenges as the provincial curriculum emphasizes discipline-centred learning and encourages age-specific achievement levels in each of the content areas. The Principal recognized this conflict noting that being a government-funded school helped financially but also created obstacles for their curriculum as, "The government curriculum and evaluations are hoops to jump while trying to retain your philosophy and approach."

Observations confirmed how the bioregional philosophy and approach that focused on holistic, integrated learning was challenged by the provincial curriculum. Although the provincial curriculum is delivered through optional workshops, it is more structured and obviously discipline-centred, particularly for the older students. This was particularly evident for science and math in the Intermediate class as these subjects were taught separately by a different teacher, replicating a mechanistic approach in practice.

The negative impact of being so discipline-centred with different teachers teaching different subjects became apparent when an opportunity for integrated teaching was missed. Although the core staff were trying to co-plan continually and co-ordinate curriculum and fieldtrip content, the senior science and math teacher planned content quite independently. This was glaringly apparent when a workshop on solar cookers was happening in the core classes upstairs while the science teacher was busy downstairs planning separate, unrelated science workshops for the afternoon. He was unaware the workshops were happening and when they were called to his attention he did not go and see what was taking place. Even though some potentially engaging math and science was being introduced and taken up by the students, it had to be dropped, as it was not incorporated into the following math or science lessons. This enacted curriculum, influenced by the provincial curriculum, was counter to the intended bioregional curriculum that stated students need not be aware of subject boundaries. As a result, the school became far more aware of the need for all teachers to collaborate in curriculum planning, incorporating this as a prerequisite for hiring new teachers.

The Hidden Curriculum

The School Manual also focuses on the "Medium" or "Hidden Curriculum." Although it relates to the school organization and administration, it is worth noting here as the school identified it as having an impact on what students are learning – overtly or covertly. By recognizing the hidden curriculum and its powerful effect, the bioregional school specifically attempted to avoid top-down control, competition and individualism, and encourage *community*, co-operation, and *interdependence* as core curricular concerns.

Curriculum Development

When asked how the curriculum was developed along bioregional lines given the fact that the school follows the provincial curriculum, a staff member recognized the process involved:

> That, and we have self-directed learning. We try to get harmony with the three of them, but it is not always possible. So it is a balancing act and sometimes we try to get to do more reading or writing or things like that. I'm trying to get them to have fun in [outdoor] sit-spots and do more awareness games and do tracking, etcetera; and then what is it that they are interested in doing? Then there is the process of integrating it. I believe we should see what their interests are and then see how to weave that in and then weave in what we learn in terms of the bioregional curriculum, for example: bringing in journal writing or a writing workshop into the experiences gained in the bioregional activities. My concern is more with the self-directedness and the bioregional curriculum. The philosophy is that they will learn way quicker and way more when they are interested in something.

Given the emphasis on self-directed learning, the Intermediate Core Teacher was asked how they integrate the Government curriculum and the Bioregional curriculum. He responded:

> In some ways they are not separate because there is a lot of overlap. We write observations on students when they do something new on

individual checklists, if they are lacking in an area we bring up specific things to address that with a specific activity. We also look at checklists of topic areas to see if we are covering those. We work on a four-year rotation in science and social studies. Some bioregional aspects are covered every day/month/year.

Rather than being taught as a specific subject, bioregional education is integrated throughout the school and curriculum in a variety of ways. When a parent and director was asked how bioregional education is incorporated, she responded,

In fieldtrips, geography, natural history; in the school culture in awareness of environmental behaviours, for example, students must take their lunch garbage home so they notice what garbage they generate; the environment and what impact they have is intrinsic in planning events.

Another parent and school director supported this notion that bioregional education is infused into the culture of the school. When she was asked if bioregionalism has a focus in the curriculum she responded,

Here is comes naturally as part of everything. It is incorporated on the fieldtrips on Fridays by learning in the environment; by the types of craft materials that are available for example: reused paper; by travelling by bikes; development of the playground as a natural environment with water being pumped up to the top with a rowboat; by a balance beam brought by bicycle.

The students also felt that bioregional education was a big part of the school curriculum in a variety of ways through how they learned outdoors and how the school modelled sustainability.

Consistently, the staff and parents confirmed the bioregional curriculum approach exemplified the ecological principles of *adaptation* and *emergence* based on research, their personal reflections, and *feedback* from students, as the curriculum has developed by experimenting with different approaches over the years. Self-direction allows the students to develop curricula, supported by teachers as mentors and facilitators; and at other times teachers develop specific workshops to address identified learning needs from the students or teachers. For example, the day tends to start with investigations that integrate math, science, and social studies, then continues with a workshop in communications to help express an outcome of the investigative learning.

To facilitate an integrated Bioregional curriculum and the Provincial curriculum, the two core teachers do a lot of joint planning and curriculum development. Given the involvement of other part-time teachers with expertise in bioregional education and volunteers, constant communication is essential and is a daily component of the organizational structure. The amount of co-operation and extensive communication was evident in the planning of social studies and science where core teachers were coplanning the year's curriculum, working closely with the greater school community at the Visioning Meeting, so that the provincial curriculum topics in science and social studies would all be addressed in a four-year cycle. This resulted in group as well as individual study options that could be used to integrate the other subject competencies through their bioregional approach.

Although the curriculum was subject centred and mechanistic at times through workshops, it was not at others when projects were developed. The follow-up visit, in October 2003 showed the new teacher initiating more planning and structure, with projects and themes being the focus of workshop options. These workshops were often connected to science and social studies themes and incorporated art, music, drama, and language. This showed an active development and belief in *emergence* in the school. The new challenge seemed to be balancing the new structures and timetabling that the new teacher introduced to help focus the older students, with a flexibility that would allow an integration of traditional subjects and self-directed learning rather than teacher-led learning. In supporting this the Principal stated, "Ideally, in addition to volunteers, we'll have two main teachers for each group: one that provides a group option and one that oversees individual initiatives and projects." Being open, based on interests, anyone can be involved in curriculum development. In addition to students and teachers, parents and community volunteers often offered learning activities.

Fieldtrips

In developing the provincial and bioregional curricula, one day a week is committed to Field Studies. The School Manual states,

> Field studies are an integral part of an experiential learning program. Weekly field trips to places of natural and cultural interest bring authentism to classroom studies. They nurture connections with our cultural and natural heritage and with the often-hidden adult world.

Fieldtrips, as a weekly aspect of the curriculum, ensure students are learning outside every week, developing ecological intelligence. Having them every Friday makes preparation from home easier; accommodates students who come only on certain days; and ensures they get out every week, so the curriculum has a strong ecological context. A teacher explained the significance of these fieldtrips,

> Part of bioregionalism is learning about place so a big part of that is learning to read the landscape. It is one thing to read the alphabet and books but there are very few people left on this planet of humans who know how to read the landscape. People are so unaware of all that is around them that it is really hard to read the landscape anymore. That's why we have the ecological and social problems today. It is trying to develop ecological literacy.

Accordingly, fieldtrips provide more integrated options for teaching science or social studies in a bioregional context. As the Bioregional curriculum is not clearly defined and laid out, topics or themes for the fieldtrips come from the provincial curriculum in science or social studies. Fieldtrips, then, are either Nature or culturally oriented but try to be interdisciplinary by offering a more integrated mode of learning. From September 2002, the core teachers were working together with the fieldtrip co-ordinator, to link the fieldtrip content to the classroom learning with cross planning between science and social studies.

Within each fieldtrip there are two required workshops in the morning that may involve sensory experiences or wilderness skills as well as optional workshops in the afternoon that develop as an *emergent* curriculum from earlier experiences and interests or from a theme of Nature Explorers that may involve mapping, tracking or bird language for example. These workshops would be more in-depth than the more general required workshops. If they choose, the students can opt for free time to explore on their own.

Ecological Intelligence

As bioregional education seems to be integrated as a theme into all aspects of the school and curriculum, questions relating directly to ecological intelligence were asked to determine the extent of its incorporation according to the more tangible terms identified by Orr (1992) such as biophilia (love of Nature), developing a land ethic, and understanding ecological principles. One of the school founders and one of the principle bioregional educators on fieldtrips was asked what he considers to be the ecological intelligence that is fostered through the school. He responded,

> It is connected to ecological literacy and relates to the physical, mental, emotional and spiritual dimensions of the person. It is the ability to respond appropriately. If we are getting feedback that there is climate change, loss of biodiversity, loss of soil, air pollution and we

are not doing anything about it, in terms of ecological intelligence we are pretty unintelligent. The ability to read the landscape and to have that connection with the landscape in all those different realms would show ecological intelligence.

A learning assistant teacher felt they were developing biophilia as, "They see by example and we talk about respecting the environment and being sensitive, appreciating and respecting."

Regarding the development of a land ethic a teacher noted,

Especially on the fieldtrips we are learning how different areas are sensitive so what is appropriate in each area and at what times of the year. Knowing your effect on the space that you are in is part of ecological intelligence: knowing your effect on other people, on animals and plants.

When asked whether the school incorporated ecological principles in the enacted curriculum, and if so how, one of the teachers replied,

Yes, through workshops and mentoring. We are building experiences first. The principles are being learnt more unconsciously now. There is a progression in learning, developing a sense of place. So depending on their age and how long they have been with us will influence the level of their learning. Many of the ecological principles are more intellectual concepts and I feel that we rush into wanting kids to understand the intellectual aspects of cycles or diversity without having the experience of that. So that is why I say we build from the experience of that and are seeing it over time before they are given the names.

Another teacher also felt ecological principles were an integral part of the curriculum but were not taught in a focused lesson format. He felt it was hard to identify how or when these were taught, "... because this is such an organic process at the school. All of us who interact with the kids are conscious of trying to find opportunities to talk about ecological principles as we go. The evolution of the fieldtrips is the really important part of that." Another teacher concurred with this integrative approach stating, "We don't run workshops to teach them, but we consciously bring ecological principles up to work them in, often through stories. They are threads that are woven through." This approach seemed to be effective in the experienced curriculum as interviews with the students revealed an extensive ability to give examples of the various ecological principles through numerous examples.

Although the analysis of the Bioregional curriculum is limited to documentary, interview and relatively limited observational data of the intended and enacted curriculum, it did reveal there is a strong resonance with Sterling's (2001) characterization of sustainable education, and conscious efforts to overcome the mechanistic influences of the provincial curriculum. Those mechanistic resonances appeared in the enacted curriculum when the bioregional school attempted to blend the two curricula through discipline-centred teaching, less so when the disciplines were the focus of workshops to support group and self-directed explorations. Significantly, the respondents identified the mechanistic metaphors in the provincial curriculum as obstacles and were working in an *emergent* way to overcome them through on-going curriculum development.

Ecosystem-Based Adaptation in Education Curriculum (UNEP-IEMP, 2019)

Another example of an eco-centric curriculum guide, supported by the United Nations Environment Program (UNEP), is The Ecosystem-based Adaptation in Education Curriculum Resource Guide. It uses various eco-systems as a larger organizing principle to integrate subject-based curricula to address the challenges of climate change. The curriculum guide is developed by Ecosystem-based Adaptation through South-South Cooperation (EbA South). According to the resource guide, the main objective is

> to position EbA as an approach to address the challenges related to climate variability and change globally. The reference guide is designed to support teachers and environmental educators to incorporate the key aspects of EbA into formal or non-formal education curriculum. It promotes awareness of the key role that ecosystems play for communities to adapt to climate change. This guide is designed to enable educators at the different education levels (primary, secondary, university) from diverse subject areas to introduce EbA across curriculum.
>
> (UNEP-IEMP, 2019, p. 3/4)

It is designed for decision makers in the education sector responsible for developing and implementing educational programmes and initiatives; primary, secondary, and undergraduate level teachers/youth educators; teacher training institutions; non-governmental organisations (NGOs) involved in the development and implementation of non-formal education programmes; and interested citizens, youth, and students.

Focusing on ecosystem-based adaptations in relation to climate change enables interdisciplinary systems thinking and learning at all levels so as to promote a thorough understanding of the role of ecosystems, reduce vulnerability, build resilience of both the social and ecological systems, and enable students to take action. The resource guide provides insight into how ecosystem-based adaptation can be integrated into formal and informal education and various school subjects at primary, secondary, and undergraduate levels enabling interdisciplinary teaching and learning. The resource guide includes suggestions for planning and preparing EbA curriculum and also includes field studies as well as example curriculum modules on ecosystem services, ecosystem adaptation, EbA in forest and mountain, marine and coastal, and dryland ecosystems. Through this ecosystem-based approach, ecological principles are highlighted to bring about essential systems thinking and empowerment in transitioning to sustainable development given the reality of climate change impacts that communities around the world are facing.

Conclusion

These examples show how teachers and schools can start transitioning to sustainable education. Whether a teacher is a subject specialist at high school or teacher training levels, or a teacher at elementary grades responsible for all subjects, developing curriculum based on a holistic eco-centric paradigm is possible. All it takes to get started is the interest, motivation, and openness to design decentralized curriculum that will support teaching and learning differently in developing sustainability as a frame of mind. As shown above, the ecological principles, the Eco-Centric Curriculum Design Framework, a bioregional approach, and an ecosystem-based approach can help develop transition to interdisciplinary topic-based curricula while incorporating provincially mandated competencies and traditional subject-centred learning outcomes.

Curriculum in our educational systems needs to be open to enable and encourage *adaptations* and *emergence* – that is, learning – based on *interdependence, community, diversity, cycling,* and effective *feedback.* This enables curriculum to emerge naturally rather than by way of a decontextualized, linear, predetermined, and imposed process that typically lacks local or personal relevance at particular times in a student's development. Such an emergent approach to curriculum development is dependent on teachers being learners, transitioning to becoming active partners in co-designing curriculum with colleagues and students in an iterative, adaptive design process. Once we transition to an ecologically designed curriculum, we can then support transitions in teaching and learning.

References

Benyus, J. (2014). Biomimicry in action. *TED Talk.* https://www.ted.com/talks/janine_benyus_biomimicry_in_action?language=en

Ireland, L. (2007). *Educating for the 21st century: Advancing an ecologically sustainable society* [Doctoral dissertation]. University of Stirling. STORRE. http://hdl.handle.net/1893/240

Orr, D. (1992). *Ecological literacy: Education and the transition to a postmodern world.* State University of New York Press.

Sterling, S. (2001). *Sustainable education: Re-visioning learning and change*. Green Books.

UNEP-IEMP. (2019). *Integrating ecosystem-based adaptation in education curriculum: A resource guide*. Document produced as part of the GEF-funded EbA South project. http://www.ebasouth.org/sites/default/files/attachments/Integrating %20Ecosystem-based%20Adaptation%20in%20Education%20Curriculm%20 Resource%20Guide_EbA%20South.pdf

---

Appendix

An Emergent Topic-Based Primary Unit Plan for Trees

By Alisha Underwood, Practising teacher and MA in Environmental Education and Communication, Royal Roads University

Topic: Trees and People

Ecological Principles

Interdependence: We need trees to live. There is a meaningful relationship between many First Nations people in BC and trees.

Community: The class has established a community surrounding a special tree.

Diversity: Different types of trees help humans in different ways.

Adaptation: Trees adapt to seasonal changes.

Activities

1 **Our special tree**
   Summary: Giving students the responsibility to work together and find a special tree may motivate students to observe the trees and find beauty in the forest. Observing the special tree throughout the year will bring awareness to what is happening in their local environment.

   a "Why did you choose this tree?" "How did you all decide this would be our special tree?"
   b "What do you notice about the tree?" Encourage the students to use their five senses to describe the tree. "How can you describe the tree?" "How does the tree feel/look/smell/sound?" "What should we name the tree?" Giving the tree a name personifies the tree and may help students connect to the tree and empathize with it.
   c The special tree will act as the class meeting spot throughout the year. Therefore, each time the class visits the forest, they draw their attention to the subtle changes the tree is making to adapt to the seasonal changes. Ask, "What do you notice?" "How has the tree changed?"

"Why is the tree changing?" Students can record these changes in their Nature journals.

BC Curriculum Subject Learning Outcomes

Language: L2, L5, L11, L17, L19, L20, L21

Career Education: CE2, CE3, CE5, CE14, CE15

Social Studies: SS1, SS9

Science: S1, S2, S4, S5, S6, S11, S12, S13, S15, S16, S17, S18, S21, S23, S24, S28, S29

2 Tree poetry

Summary: As an extension to "Our Special Tree" activity. The students can write a poem together about the special tree by each offering a word to describe the tree. In this way the students can experience the tree as poetry and as art.

a Position the students in diverse ways around the tree: around the trunk looking up at the tree, standing looking at the bark, sitting looking at fallen leaves, and lying down under the tree looking straight up at the canopy.

b Give the students time to quietly observe the tree from their unique vantage point.

c Explain that each student will give one word to describe the tree. From their spots beneath the tree, each student offers a word to describe the tree. The teacher records these words on a piece of paper and arranges them into a poem.

d Back in the classroom, the teacher can print out a copy of the poem for everyone to glue into their Nature journals.

BC Curriculum

Mathematics: M4

Language: L1, L2, L5, L17, L19, L21

Career Education: CE1, CE2, CE3, CE4

Arts Education: A1, A3, A4, A5, A15

Science: S1, S16, S18, S23

3 We need trees

Summary: We need trees and use trees every day. This activity can help students understand the value of trees and guide them to a place of appreciation. Trees provide everything humans need to be alive: oxygen, food, shelter, and warmth. Trees are also beneficial for students' well-being and can fill humans with positive and calm emotions.

a Gather students in the gathering place and ask, "How can we live in the forest as part of the forest?" Explain to the students that we will pretend to live in the forest. Ask, "How can trees give us everything we need?" "How can they provide us with shelter?" "How can they provide us with food?" "How can trees protect us?" "How can trees help us collect water?"

b Group students in pairs and explain that they will be making shelters using materials from the trees. Ask, "What parts of the tree can you use to build your shelter?" "How will you make sure your shelter is big enough for you to fit inside?" "Where should you build your shelter?" "Why shouldn't you build your shelter under a fir tree?" Give students plenty of time to build their shelter.

c When the students are finished building their shelters, gather the students and ask, "What will you eat?" Encourage students to find food from the trees that they can store in their shelters. Tell them not to actually eat the food, just pretend.

d Finally, ask, "How can you use the trees to collect water?" Encourage students to make something using the tree materials that will help them collect water.

e When the students are finished, they can present their shelter, food, and water to the class. Ask "How did you build your shelter?" "What materials did you use?" "What food did you find?" "How did you collect water?"

BC Curriculum

Mathematics: M4, M13, M17, M22, M25, M26, M29

Language: L1, L2, L5, L10, L17, L19, L20, L21

Career Education: CE1, CE2, CE3, CE4, CE5, CE6, CE8, CE14, CE15

Arts Education: A2, A13, A15

Applied Design, Skills, and Technologies: ADST1, ADST2, ADST3, ADST4, ADST5, ADST6, ADST7, ADST8, ADST9, ADST10, ADST14, ADST15, ADST16, ADST17

Social Studies: SS9, SS11, SS13

Science: S2, S4, S11, S12, S15, S19, S20, S21, S24

Physical and Health Education: PE5, PE12, PE13, PE16, PE18, PE25, PE26

4 Whittling wood

Summary: Whittling with wood allows students to explore the properties of the tree and its wood by scraping off the different layers to explore what is underneath. This activity was inspired by Dr. Enid Elliot who has experience whittling wood using a vegetable peeler with her forest kindergarteners.

a Gather students outside around a blanket. Show students various wooden artefacts and lay each artefact out on the blanket for everyone to see. If you do not have access to wooden artefacts, you can show the students images of wooden artefacts.

b Explain to the students that they will be whittling wood by using a vegetable peeler. Ask safety-related questions, "How can we stay safe using a vegetable peeler?" "Which direction should we peel?" "Why is it important to peel away from your body?" Demonstrate how to whittle wood using a vegetable peeler by slowly pushing the vegetable peeler away from your body and only slicing off a tiny piece of the wood each time.

c Lead the class on a Nature walk to collect sticks. Remind students to choose sticks that are smooth and not too dried out.

d Work with small groups of students to help them safely whittle their stick. While they are whittling ask, "What do you notice about the wood under the bark?" "How does it feel?" "What does it smell like?" "How is it different from the bark?" "How is the stick changing?"

e Once everyone has had a turn to whittle, gather the group in a circle with their sticks. Encourage the students to use their imaginations to tell everyone what they made with their sticks.

BC Curriculum

Mathematics: M4, M13

Language: L2, L5, L17, L19, L20, L21

Career Education: CE4, CE5, CE6, CE8, CE12

Arts Education: A1, A2, A5, A10, A13, A14

Applied Design, Skills, and Technologies: ADST1, ADST2, ADST3, ADST8, ADST15 ADST17

Science: S1, S4, S8, S9, S10, S15, S16, S17, S18, S19, S20, S21, S23, S28

5 The Tree of Life

Summary: To many First Nations people, the cedar tree is known as the "tree of life" because it is used for shelter, clothing, totem poles, masks, tools, bracelets, hats, canoes, boxes and for spiritual cleansing. I will invite an elder or First Nations Educator to lead this activity or an activity related to the cedar tree, so students understand the different uses of the cedar tree and its importance to many First Nations people.

a Gather students outside around a large cedar tree and read the book *Cedar – The Tree of Life* by Boreham (2013). This book explains some of the ways that the cedar is useful to the people of the Pacific North-west Coast.

b After the story, ask, "Why is the cedar tree important to many First Nations people?" Record their ideas on a piece of paper to display in the classroom.

c Explain that First Nations people "give thanks" when they take some-thing from Nature to show their appreciation and gratitude.

d Guide students to find something in Nature to be thankful for. Rec-ognizing elements in the natural world that they are thankful for may deepen the students' understanding that they are part of the inter-connected ecosystem and therefore they may begin to more deeply appreciate the natural things that help keep us alive.

BC Curriculum

Language: L2, L4, L5, L8, L9, L10, L13, L15, L16, L19, L20, L21

Career Education: CE1, CE2, CE3, CE4, CE5, CE9, CE13, CE14, CE15

Social Studies: SS1, SS2, SS4, SS5, SS8, SS9, SS14

Science: S7, S14, S22, S29

Physical Education: PE27

Topic: Trees in the Elements

Ecological Principles

Interdependence: Trees are part of an interdependent ecosystem and rely on plants, animals, and the elements to survive. Life is dependent on trees because trees make oxygen.

Community: A tree is never alone; trees live in a community of beings that depend on each other.

Diversity: A variety of trees in the environment can help keep the forest healthy.

Cycling & adaptation: Trees change throughout the seasons to adapt to the weather and temperature.

Activities

1 **Draw a tree**
Summary: This activity was inspired by Dr. Liza Ireland and shows the interconnectedness of people, plants, animals, elements, and trees. By drawing a tree as a class, students will see how people, plants, animals, and elements depend on trees as well as the needs of a tree. This activity

illustrates that a tree is not an individual entity but lives in a community and relies on its interdependent relationships for survival.

a Gather students on the carpet in the classroom in front of a whiteboard. Explain to the class that you need their help to draw a tree. Have students give ideas of what to draw, for example, roots, soil, leaves, the sun, rain, clouds, a swing, squirrels, and a nest. Draw all their ideas on the whiteboard. Continue to prompt students to give more ideas of what to draw by asking, "Anything else?"

b Once finished, discuss the picture and ask, "What do you notice about the tree?" "What does the tree need to survive?" "Why did we draw a _____?"

c As an extension, students can draw trees outside in their own Nature journals thinking about all the interdependent relationships that a tree has.

BC Curriculum

Language: L1, L2, L5, L10, L17, L19, L21
Career Education: CE2, CE3, CE14
Arts Education: A4, A14, A18, A20
Science: S1, S4, S5, S6, S11, S12, S13, S18, S21, S23, S24, S28

2 **Pretend to be a tree**
Summary: By embodying a tree who is rooted in place throughout different seasons and weather, this activity may help student develop a sense of empathy for trees. The students can gain the perspective of what it is like to be an other-than-human being who must adapt to the elements.

a Gather students around the special tree and ask, "How does the tree feel?" "How do you think the tree feels in the rain/snow/wind/sun/ Fall/Winter/Spring/Summer?"

b Explain that they will pretend to be a tree. The teacher can show the students how to stand tall like a tree and how to extend their arms up to the sky to gather energy from the sun and stretch their toes to feel the earth beneath their feet. Remind the students that a tree cannot move from its spot because it is rooted in place.

c Guide students through different weather and seasons, for example, "It is Spring. All your leaves are green, and you are a tree in the sun." "It is Summer. It is very hot, and you are a thirsty tree. Your leaves are turning brown." "It is Fall. You are a tree, bending, swaying and losing your leaves in a windstorm." "It is Winter. You are a tree who has lost all their leaves and is now covered in snow."

d Since the students are pretending to be a tree experiencing different elements, have students reflect on how they felt as the tree and ask, "As a tree, how did you feel when it was hot/cold/rainy?"

BC Curriculum

Language: L5, L10, L17, L19, L21

Career Education: CE1, CE3, CE4, CE6, CE14

Arts Education: A2, A3, A5, A13, A20

Science: S1, S4, S5, S6, S11, S12, S13, S15, S16, S17, S18, S21, S29

Physical and Health Education: PE1, PE5, PE6, PE16, PE25

3 **Using our five senses, observe how trees change throughout the seasons**
Summary: To connect with a tree and to observe its natural seasonal changes, have students record a special tree in their Nature journal on a daily or weekly basis. Students are encouraged to use all their senses to record the tree so they can understand it more deeply. Their Nature journal will be authentic evidence of how the tree adapts and changes with the seasons.

a Gather students in the designated forest gathering place. The teacher explains that in their Nature journals they will write about or draw a tree. Explain that the tree will become their "special tree" because they will visit this tree on a regular basis to observe and journal with it.

b The teacher can model how to set up the journal page and show the students how to carefully observe what the tree looks like using all their senses. The teacher can model how to observe the tree by thinking out loud and asking, "What colour are the leaves?" "What do the branches look like?" "Is the trunk covered is moss?" "What does the moss feel like?" "What does the bark feel like?" "What do the leaves smell like?" "What do the seeds, cones and nuts on the ground around the tree look like?" "What does the tree sound like?" "How can you describe the size of the tree?"

c The students then get their journals and writing tools from their backpacks and find a sit spot at the base of a special tree to write and draw in their journals. Even if they are finished, they must stay in their sit spot to connect with the natural world around them.

d Once finished, they can share their trees with the class. Since each student will have observed a different tree, sharing their journals may help students see the diversity of trees in the forest.

BC Curriculum

Mathematics: M1, M14, M18, M26, M29

Language: L1, L2, L3, L5, L6, L10, L11, L17, L19, L20, L21

Career Education: CE1, CE4, CE14

Arts Education: A1, A2, A5, A13, A15, A20, A23

Social Studies: SS9, SS13

Science: S2, S11, S12, S23, S24

4 Trees make oxygen science experiment

Summary: Trees make oxygen that sustains life. This science experiment shows that when submerged in water, oxygen bubbles cling to the needles of coniferous trees. Therefore, students can see the tree's needles producing the oxygen that they breathe.

a As a class go on a Nature walk to observe the coniferous trees in the local area. Explain that the students will know the tree is coniferous because coniferous trees stay green all year and don't lose their leaves in the winter. When you find a coniferous tree, such as a cedar, pine, and fir, introduce the students to the tree and take a small cutting of the needles for the science experiment. The teacher can pass around the tree cutting for students to feel and smell.

b While walking back to the classroom, the teacher asks, "How do trees breathe?" "How do trees help us breathe?" "How do tree clean the air?" "Why are these needles important to the tree?"

c Once in the classroom gather on the carpet and ask, "What do you think will happen if we put these needles in water?" Record the students' answers on the whiteboard.

d Label the jars according to the different trees (cedar, fir and pine) and submerge the needles in water. Let the needles sit in the water for an hour.

e After an hour, observe which needles produced the most bubbles. You may be able to count the bubble on the needles. Explain that the bubbles on the needles are oxygen from the trees. Ask, "Why do you think there are oxygen bubbles on the needles?" "Why do you think this tree's needles produced the most oxygen bubbles?"

BC Curriculum

Mathematics: M1, M4, M6, M11, m18, m19, M24, M27

Language: L2, L5, L6, L10, L17, L19, L21

Career Education: CE1, CE4, CE14

Applied Design, Skills, and Technologies: ADST1, ADST16

Science: S1, S2, S5, S6, S12, S13, S15, S18, S21, S23

Topic: A Tree's Life

Ecological Principles

Interdependence: Trees live in interdependent relationships with other trees by communication with each other through mycelium networks.

Community: Trees use mycelium to share water and nutrients.

Diversity: A variety of trees in the local environment can help keep the forest healthy.

Cycling: The lifecycle of a tree. All stages of a tree's life support the forest.

Activities

1 **Create a master list of trees in the local area**
   Summary: Have students take pictures of trees in the local area and create a master list of local trees. Taking pictures of trees and observing their features can help introduce students to the diversity of the forest as well as teach them which trees are deciduous and coniferous. Students may have done a similar activity already in a previous unit, so they may already know how to use a camera or an iPad to take pictures.

   a Before starting the activity, make sure each pair of students has a camera or an iPad to take pictures. Meet outside in the gathering place. Explain that the students will be searching for different kinds of trees in the forest area and photographing them. Ask, "How many trees grow in this forest?" "How can we tell if trees are different or the same?" "Do all trees grow in the same place?" Have students find a partner and give each pair a digital camera or an iPad.
   b Encourage the students to look closely for different trees in the forest area. Prompt the students to observe the tree's bark, leaves, size, shape and colour.
   c After the students have taken pictures of the trees, print the pictures and have students compare their images to other groups' images and to field guides to discover the tree's name and features of the tree. Ask students, "How can this tree be used?" "How is the cedar tree different from the Douglas fir?" "Where do the tallest trees grow?" Encourage students to pay attention to various characteristics of the trees such as the bark, leaves, size, colour and shape.
   d As a class, create a master list of trees in the local area.

BC Curriculum

Mathematics: M1, M4, M13, M18
Language: L1, L2, L5, L17, L19, L21

Career Education: CE2, CE3, CE14, CE15

Arts Education: A2, A3, A4, A5, A13, A14, A15, A22

Science: S1, S4, S5, S6, S8, S15, S16, S18, S20, S21, S23, S28

2 Observe various deciduous and coniferous trees and have students draw the trees in their Nature journals

Summary: In their Nature journals encourage students to carefully observe and draw deciduous and coniferous trees in the forest. This journaling activity may help students notice the differences between trees such as colour, shape, and size.

a With their Nature journal, have students gather around a coniferous tree such as a pine, cedar, or fir. Ask, "What do you notice about this tree?" Show students how to draw the triangular shape of the coniferous tree, how the trunk grows up from the base of the ground, and how the roots and mycelium network extend from the base of the tree to other trees in the forest so that all are connected underground, forming a community.

b Give students time to draw the tree. Once they are finished, students sit quietly in Nature to observe and connect with their surrounding environment.

c If the students are engaged, continue by doing the same activity with a deciduous tree. If they are not engaged, students can draw a coniferous tree on another day.

d Once they have finished drawing both trees encourage students to compare them by asking, "What is the same?" "What is different?"

BC Curriculum

Mathematics: M4, M13, M14, M29

Language: L2, L5, L10, L17, L19, L21

Arts Education: A1, A2, A3, A4, A5, A20, A22, A23

Social Studies: SS9, SS11

Science: S1, S4, S5, S6, S8, S11, S13, S15, S16, S17, S18, S20, S21, S23, S24

3 Lifecycle of a tree scavenger hunt

Summary: This lifecycle scavenger hunt can show students the various stages of the lifecycle of a tree: seed, sprout, sapling, mature, snag, and rotting log. This scavenger hunt allows students to authentically discover the various stages in a tree's life and use all their senses to carefully observe what happens to the tree during each stage of its life.

a Prepare for this activity by creating a worksheet that shows the lifecycle of a tree. The students can use this worksheet as a guide when they are searching the forest. Also, bring in a large tree cookie that shows the growth rings of the tree.

b Meet outside in the gathering place and show the students the lifecycle worksheet and ask, "How can we tell how old a tree is?" Explain the various stages of a tree's lifecycle.

c Start the scavenger hunt as a group and lead the class to a very large tree, and ask, "What can the size, of the tree tell us?"

d Next lead the class to a tree stump and ask, "What do the rings of the tree tell us?" As a class, count the tree's growth rings. If there is not a stump in the local forest, show the students a tree cookie. Ask, "What can the circumference of the tree's trunk tell us?" If the students seem interested, give them a piece of string to use to measure the circumference of trees in the forest to determine which tree is the oldest. The oldest tree is the "mother tree" and takes care of all the other trees in the area.

e After, lead the class to a dead tree or a rotting log and ask, "How do dead trees help the forest?"

f Now, with their partners, the students can search the local forest for different stages of the lifecycle of a tree. Give students plenty of time to explore the forest to search for the lifecycle of trees.

g When they're finished, ask, "Did you discover anything interesting that you would like to show the class?" Give students the opportunity to explain and lead the class to their discoveries.

BC Curriculum

Mathematics: M1, M4, M13, M18

Language: L2, L5, L10, L11, L17, L19, L21

Career Education: CE2

Social Studies: SS9, SS11

Science: S1, S5, S6, S15, S16, S17, S18, S21, S23, S25, S28

4 Find the tree's mycelium network

Summary: Trees communicate with each other by using an underground network made of fungi that grows at the roots of trees called the mycelium network. In this activity students will observe the little white threads of the mycelium network.

a In preparation for this activity, you will need a shovel and a way to show the students a video. Gather students together in the classroom and ask, "How do trees talk to each other?" Explain the mycelium network and show the class a video of how the system works and pictures of a forest's mycelium networks.

b Lead the class on a walk outside and remind the students that trees talk to each other using their mycelium network. Ask, "What do trees talk about?"

c On the walk, guide the class outside to the largest tree in the forest. Explain that this tree is the "mother tree" and she uses the mycelium network to talk to the other trees and to help them. Ask, "What do you notice about the mother tree?" "How do you think the mother tree helps the other trees?"

d Taking extra care to ensure the tree is not harmed, use your shovel to dig down to the mother tree's mycelium network (the little white treads that go everywhere). Ask, "What do you notice?" "How does the mycelium network help all the trees in the forest?" "What happens if we get sick?" "What happens if a tree gets sick or hungry?"

e Students can document their observations in their Nature journals

BC Curriculum

Language: L1, L2, L5, L6, L10, L11, L17, L19, L21

Career Education: CE2, CE14, CE15

Social Studies: SS1, SS9, SS11, SS13

Science: S1, S4, S5, S6, S13, S15, S16, S17, S18, S19, S20, S21, S23, S24

5 **Play games that encourage students to closely encounter the local trees**
Summary: Play games that encourage students to build relationships with the local trees. The following games, Meet-A-Tree and Tree Tag, are from the book *Coyote's Guide to Connecting with Nature* (Young et al., 2010) and they can help students discover, understand, and appreciate trees.

Meet-A-Tree game (Young et al., 2010, p. 456). The Meet-A-Tree game helps students expand their senses by use senses other than sight, such as touch, smell, and hearing to explore the details of a tree: branches, bark, roots, and moss growth. This activity gives students time to explore a tree in ways they may not have before and therefore they may notice something new and different.

a Have students find a partner and each pair is given a blindfold. The pairs decide who will be blindfolded first.

b The blindfolded person is gently guided by their partner to a tree. Once at the tree, the blindfolded person gets to know the tree using all their senses other than their eyes. Encourage the blindfolded person to hug the tree, smell its branches, rub their skin against its textures, feel the roots and the circumference of the tree.

c When finished, they are led back to the starting point. Then they remove their blindfold and need to find the tree.

d After the first person successfully finds their tree, switch partners so they can have a turn.

e As an extension, the teacher can lead the group around the forest to meet all the trees. The students can introduce their trees to the class and introduce their trees as new friends and tell something special about their trees.

Tree Tag game (Young et al., 2010, p. 459). This game helps students recognize the different trees in the forest and notices the features and diversity of trees. This is a fun and active way for students to learn how to identify trees.

a Gather students in the gathering place and explain how to play the Tree Tag game. The chaser will shout, "You're safe if you are touching an evergreen/deciduous/cedar tree." The students must run to and touch this tree to be "safe." If a student is touching the wrong tree, the chaser can tag this student, and if caught, they will help the chaser in the next round.

b To scaffold the students' learning, show students some pictures of trees from field guides and go on a walk around the playing boundary to help students identify the different trees they can touch.

BC Curriculum

Mathematics: M4, M13, M17

Language: L2, L5, L17, L19, L21

Science: S1, S2, S4, S5, S6, S12, S13, S15, S16, S17, S18, S21, S28

Physical Education: PE5, PE6, PE7, PE16, PE18, PE19, PE25

Topic: Trees and Animals Living Together

Ecological Principles

Interdependence: Trees and animals have a symbiotic relationship because animals help new trees grow by dispersing seeds; and animals depend on trees for survival. Trees provide oxygen and absorb carbon dioxide.

Community: The trees in the local area are influenced by animal activity and interactions.

Diversity: Trees are used by a variety of animals and each animal uses trees in a different and unique way. There are many types of trees.

Adaptation: Animals plan ahead and change their behaviour in preparation for the changing seasons. Trees adapt to the changing seasons.

Activities

1 **Scavenger hunt for signs of animals**
   Summary: Searching for signs of animals in the form of a scavenger hunt can be a fun way to introduce students to the various animals that rely on trees for their survival. This scavenger hunt can help students discover that animals use trees for shelter, food, as a perch for hunting and a way to communicate with other animals.

   a Prepare for this activity by creating a worksheet with pictures and names of local animals that rely on trees such as birds, racoons, beavers, bears, and squirrels. Students can check off the animals when they find signs of them.
   b Gather students in the gathering place, beneath the special tree and ask, "Why are trees important to animals?" "How can you tell if an animal has used this tree for food, shelter, a perch to hunt or to communicate with other animals?" Record students' ideas on a piece of paper that can be posted in the classroom.
   c Give each pair of students a worksheet and explain that they will be searching for signs of animals on trees, in trees and under trees. Give students lots of time to explore the trees in the area for signs of animals.
   d When students are finished, gather the class at the special tree and ask, "Did you discover anything interesting that you would like to show the class?" "Where did you discover signs of animals?" "How do you think the tree is affected by the animals?" Give students the opportunity to explain and lead the class to their discoveries.

BC Curriculum

Mathematics: M4, M13
Language: L1, L2, L5, L17, L19, L21
Career Education: CE2, CE3, CE4, CE5, CE14, CE15
Social Studies: SS9, SS11
Science: S1, S5, S6, S13, S15, S17, S18, S21, S23, S28

2 **Build a bird nest**
   Summary: In this activity, students will build their own bird nests. By building a nest using natural materials found in the forest, students may better understand that birds rely on trees for shelter.

   a Prepare for this activity by printing out pictures of bird nests or by bringing in a real bird nest to show the students. Meet outside at the gathering place and ask, "Why are trees important to birds?" "What do birds use to build their nests?"

b Lead the class on a Nature walk to look for bird nests in the surrounding trees and ask, "Where do bird build their nests?" "Why do you think birds build their nests high in trees?" "Why is it tricky to spot bird nests?" "How do you think birds build their nests?"

c Gather the class and explain that they will be building their own bird nests using materials they find in the forest. Have students work in pairs to collect natural materials such as leaves, twigs, and feathers to build their own bird nests.

d Give students plenty of time to build their nests and collect materials as needed because they will probably need more twigs, leaves, and feathers than they initially thought. Bring in some fake bird eggs so the students can make their nests the right size and shape to keep the eggs safe.

e When the students have completed their nests, have groups share their nests. Ask inquiry questions: "How was building the bird nest?" "How do you think birds build their nests all by themselves?" "How will your nest keep the birds safe?"

f Show students the nest built by birds and ask, "How does this bird nest differ from the ones we made?" "How do birds build their nests without hands?"

g Take pictures of the nests to display in the classroom or have students draw their nests in their Nature journals. Leave the nests in the forest for birds to use and let the materials return to the forest.

BC Curriculum

Mathematics: M18, M22, M24, M26, M29

Language: L1, L2, L5, L10, L17, L19, L20, L21

Career Education: CE1, CE2, CE3, CE4, CE5, CE6, CE8, CE14, CE15

Arts Education: A2, A13, A15

Applied Design, Skills, and Technologies: ADST1, ADST2, ADST5, ADST6, ADST7, ADST8, ADST9, ADST10, ADST14, ADST15, ADST16

Social Studies: SS9, SS11

Science: S2, S15, S19, S20, S21, S24

3 **Eagle Eyes**
Summary: This predator and prey game from the book *Coyote's Guide to Connecting with Nature* (Young et al., 2010, p. 258) can show students how birds use trees for hunting. Students will understand the different ways animals use trees and they may empathize with birds and understand how difficult it is for birds to find food to eat.

a This game is best played in a forested area with places to hide such as bushes, trees, rocks, and tall grass. Before playing the game, help students check the area for hazards such as poison ivy and wasp nests.

b Explain to the class that one person is the Eagle who will stand and stay in the "Eagle Nest" for the whole game. The "Eagle Nest" should be higher than the rest of the forested area such as a stump or small hill. The Eagle closes his/her eyes and counts to 60 (with the teacher's support) whilst everyone else hides in a broad circle around the Eagle Nest.

c The hiders pretend to be the Eagle's prey such as mice, voles, and rabbits. They must hide in such a way that they can see the Eagle with at least one eye at all times. They must also hide within the determined boundary.

d The objective for the hiders is to move as close to the Eagle without being seen.

e The Eagle opens his/her eyes and looks and listens all around for everyone hiding. When the Eagle spots someone hiding, the Eagle must describe what the person hiding is wearing and point to their exact location. That person comes to the Eagle Nest and sits down.

f When the Eagle cannot see any more people, he/she closes his/her eyes again and counts to 30 while everyone quickly hides again moving at least five steps closer to the Eagle Nest. Keep playing until everyone is found or until one person remains. The last person hiding can give a bird call to help locate them.

BC Curriculum

Mathematics: M6, M27

Language: L2, L5, L10, L17, L19, L21

Career Education: C4, C6

Science: S1, S5, S6, S18, S21, S23, S28

Physical and Health Education: PE1, PE6, PE7, PE18, PE19

4 Hide and Seek Squirrel Game

Summary: This game has been adapted from *Whales' Natural Resources Activity and Games Guide* (Natural Resources Whales, n.d., p. 1) and is intended to show students the reciprocal relationship between trees and squirrels. This game shows students that squirrels eat seeds and how seeds are dispersed by squirrels.

a Gather students in the gathering place and ask, "Why do squirrels need trees?" "Why do trees need squirrels?" Explain that one of the reasons trees need squirrels is to help more trees grow. During the winter squirrels can't find enough food to eat, so in the fall squirrels plan ahead by collecting nuts and seeds and burying them to eat later. When squirrels forget where they buried their food, a new tree grows.

b   Show examples of some local squirrel food: acorns, maple seeds, and pinecones. As a class go on a Nature walk to find squirrel food. Have each student collect six pieces of squirrel food. When each student has six pieces of squirrel food, return to the gathering place and explain the Hide and Seek Squirrel Game.

c   Explain that is it currently the month of November and the squirrels need to hide their food in preparation for winter. Give students time to hide their food – each piece of food must be hidden in a different place. The squirrels need to do this quickly and must return to the gathering place by the time you count down from ten and yell "It's December."

d   Once gathered explain that it is now the month of December and the snow is covering the ground. At the teacher's signal, all the squirrels need to find two pieces of food and return to the gathering place before the teacher counts down from ten and yells "It's January." Squirrels who do not return back in time can be eliminated from the game or can continue to play for fun. Play two more rounds of the game yelling "It's February" and "It's March" to gather the squirrels. The teacher can ask mathematical questions during this game, such as "You have 2 seeds, how many more seeds do you need to find to make 6?"

e   After the game ask, "What are some reasons squirrels may not have found all of their hidden food?" "What happens to seeds that are hidden for the winter and not eaten?" "As a squirrel, how will you hide your food next time we play the game?"

f   Play multiple rounds of the game so students can develop their hiding and seeking strategies.

BC Curriculum

Mathematics: M1, M2, M19, M21, M22, M23, M24, M26, M27
Language: L2, L5, L10, L17, L19, L20, L21
Science: S1, S4, S5, S6, S11, S12, S13, S18, S28
Physical and Health Education: PE1, PE6, PE7, PE16, PE18, PE19

5   **Observe and record which animal visits a tree**
Summary: In their Nature journals, encourage students to record which animals they see visit a tree. Recording these animals can provide students with an authentic record of the animals that live in and rely on the local trees for survival. Journaling animals in trees can give students a deeper insight and understanding of the importance of trees to animals. This activity

can be done multiple times throughout the school year to help students notice how animals in trees adapt based on the weather and seasons.

a Before gathering the children at the designated gathering place, set out sit spot "lily pads" at the base of trees.
b Gather students at the gathering place and explain that in their Nature journals they will draw or write about animals they see in trees. The teacher can model how to draw and write in the natural journal and how to set up the page with the date, weather, and temperature in the upper right-hand corner.
c The students then get their Nature journals and writing tools from their backpacks and find a sit spot "lily pad" to observe and document the animals they see in trees. Even if they are finished writing and drawing in their Nature journals, they must say in their sit spots to connect with the natural world around them.

BC Curriculum

Language: L1, L2, L3, L5, L6, L10, L11, L17, L19, L20
Career Education: CE1, CE4
Arts Education: A1, A2, A5, A13, A15, A20, A23
Social Studies: SS9
Science: S2, S4, S11, S12, S13, S23, S24

6 **Animals on totem poles**
Summary: This activity is based on the children's book *Six Cedar Tree Animals* by Landahi and Aleck (2017) about a wise Eagle learning from the cedar trees. The animal symbolism in the story teaches students how many Indigenous peoples have lived with the land and have a relationship with all living things.

a Gather students around the base of a cedar tree and read the beginning of the book about the Eagle flying in the sky and resting in a cedar tree. Ask, "What can the eagle learn from the cedar trees?" This book is long for kindergarteners, so teachers may choose to read the book over several days. One day dedicated to each animal: Eagle, wolf, raven, salmon, bear, beaver, and orca. After each section ask, "What can the ____ learn from the cedar trees?"
b Each animal in the story symbolizes and represents human characteristics and qualities that students can connect with. For example, wolf is a good communicator, raven is a creative thinker, salmon is a careful and curious thinker, bear shows respect for self and others, beaver cares for the community and the environment, and orcas know about

themselves. Ask, "How are you like _____?" "Which animals do you identify with the most?"

c After the story, explain that many Indigenous people make totem poles out of cedar trees and totem poles often represent their family and are used to tell a story. Show students images of totem poles and help them identify which animals they see. Ask, "What story is this totem pole telling us?" "What animals would you use to describe you and what you'd like to be?"

BC Curriculum

Language: L1, L2, L4, L5, L8, L9, L10, L11, L13, L15, L16, L17, L19, L21
Career Education: CE1, CE3, CE4, CE5, CE9, CE12, CE14
Social Studies: SS1, SS2, SS4, SS8, SS10, SS14
Science: SS17, SS22, SS28, SS29

Unit Assessment

As in the first unit about safety, teachers will continue to document their observations in the students' personal files and indicate which of the aforementioned curricular competencies have been achieved and to what degree according to their school's reporting policy (beginning, developing, applying, or extending). In this way the teacher will be aware of which curricular competencies have been covered and which curricular competencies still need to be addressed. The teacher can use this information to ensure the missing competencies are evaluated in subsequent activities and units through well-rounded inquiry that enables the exploration of safety through multiple subject disciplines and lenses.

In addition to the evaluation of the BC curricular competencies through the documentation of observations, teachers will record anecdotes and take pictures of students exemplifying the ecological principles (*interdependence, community, diversity, cycling, adapting, emerging,* and *the whole is greater than the sum of its parts*). Pictures will be used as "evidence" of learning and displayed in the classroom and sent home to parents. The ecological principles in the unit on trees will be assessed in the following way:
Interdependence:

- Understanding the symbiotic relationship of trees, animals, plants, elements and humans
- Recognizing that living things depend on trees for survival
- Respecting the relationship between many First Nations people in BC and trees

Community:

- Understanding that trees live in a community of beings that depend on each other for survival
- Noticing that trees are influenced by animal activity
- Thanking Nature for keeping us alive

Diversity:

- Recognizing that different trees help humans and animals in different ways
- Noticing the different resources trees provide that are used by animals and humans such as food, shelter, and medicine.
- Noticing the diversity of trees in the local environment

Cycling & adaption:

- Understanding that trees change their appearance and behaviour depending on the weather and seasons
- Understanding that all stages of a tree's lifecycle support the forest
- Recognizing how animals rely on trees for survival during the different seasons

The whole is greater:

- Understanding that trees live in a community and rely on each other for survival
- Noticing that the features (truck, bark, leaves, branches) of the tree protect the tree and help it grow

References

Boreham, B. (2013) *Cedar - The tree of life*. Strong Nations Publishing Incorporated.

Landahi, M. & Aleck, C. (2017). *Six cedar tree animals*. Strong Nations Publishing Incorporated.

Natural Resources Whales. (n.d.). Retrieved April 29, 2021, from https://cdn.cyfoethnaturiol.cymru/media/686793/activities-and-games-seed-dispersal.pdf

Young, J., Haas, E., & McGown, E. (2010). *Coyote's guide to connecting with nature* (2nd ed.). OWLLink Media.

11 Transitioning Teaching and Learning

When considering how we can transition teaching and learning, it is important to recognize how the dominant mechanistic paradigm is influencing our practice and then map out ways problematic aspects can be addressed. As noted in previous chapters, professional development for teacher trainers as well as K–12 teachers is important to help teachers identify areas of their practice that can be improved to transition from mechanistic to eco-centric teaching and learning in developing sustainable education. As Sterling (2001) states, "Ecological thinking entails a shift of emphasis from relationships based on separation, control and manipulation towards those based on participation, empowerment and self-organization" (p. 129). Being a reflective practitioner is key, willing to adventure into the unknown, taking risks and letting go of control with an open attitude to learn, accepting it will be an iterative process of learning what works, what doesn't, and then finding ways to *adapt* to enable sustainable teaching and learning to *emerge*. Throughout this transition, the ecological principles provide an excellent framework to guide the process.

How Ecological Principles Guide Our Transition

We've looked at how ecological principles, as the foundation of the whole educational system, inspire, lead to, and support sustainable education; and, in Chapter 7, what that means in particular for teaching and learning. Now let's look at how we can use these same ecological principles to set in motion next steps in letting go of industrial schooling practices in how we teach and learn to embrace the opportunities before us.

Interdependence

Interdependence guides us to focus on the interdependent relationships between teachers and students, parents, colleagues, administration, the community, and the more-than-human world. Strengthening and developing

DOI: 10.4324/9781003389590-15

reciprocal relationships is a great start to transition away from top-down centralized control and teaching in silos.

Starting with students and parents, inviting them to be part of collaborative planning with the teacher, with options as to when, where, and how learning happens, can open new doors to transitioning. Asking about students' interests and needs is an important step in developing collaboration. Teachers typically communicate well with parents telling them what the class is doing and how well their child is learning, but expanding this relationship to be a reciprocal partnership with students and parents will strengthen teaching and learning opportunities. To enable collaboration, transparency is key. Students, parents, and administration need to be brought into the process to gain their support, helping them understand new ways of teaching and learning are being implemented as adaptive learning opportunities in order to help students develop the competencies needed to thrive in transitioning to a sustainable society.

As noted in Chapter 7, it is extremely important to teach and learn in ways that reinforce our interdependent relationship with our natural environment. Easy ways to get started include learning and working in school gardens, sitting quietly with students in the natural environment being place-responsive, drawing or writing in a special outdoor spot, or engaging in various activities in diverse outdoor environments, learning from Nature. In this way, the whole of the student and the teacher, their minds, bodies, and the environment, are interdependently contributing to their understanding of the systems we are integrally a part of.

Community

In branching out into new pedagogical approaches, it's important for teachers to develop a strong professional learning *community*. Developing and working with administrative support can help teachers feel free to try new approaches in a "safe-to-fail" learning cycle. In-service professional development at the school or district level often involves professional learning communities that self-organize around particular areas of interest. These are excellent opportunities to collaborate and develop interdisciplinary transition initiatives and have a community of practitioners that include administrators, parents, community members, post-secondary education students, and post-secondary instructors as well as teachers to support each other and extend the learning. This learning community also needs to involve Teachers On Call as they are often overlooked yet are essential members of the teaching community, standing in to teach a variety of courses and levels at a moment's notice.

At teacher training institutes, high schools, and at the elementary level where there are separate subject specialists, there are opportunities to collaborate with colleagues in developing systems thinking through inter-disciplinary teaching and learning initiatives. This can include multi-age collaborative learning opportunities with older students learning with and at times mentoring younger students. Systems thinking helps recognize *interdependence* and strengthen teaching and learning in *community*.

Teacher training institutes are perfectly placed to integrate this supportive community approach into teaching practicums so as to initiate and support transitions in teaching. School practicums offer student teachers opportunities to try out sustainable education teaching methodologies with the support of their college/university instructors as well as the classroom teacher they are working with. In this way, mentoring teachers may be introduced to eco-centric teaching and learning, and conversely, mentoring teachers can help student teachers gain the insights and support needed in translating theory into practice. Teachers on call also have an opportunity to introduce transitional practices when working with various teachers, offering alternative teaching methods to the network of schools, teachers, and students they work with.

Diversity

Considering how to incorporate diversity brings up further opportunities to help transition teaching and learning. As we saw, strengthening interdependent relationships with students, parents, administration, and the community opens the door to enable and address *diversity*. A teacher can bring in diversity and different ways of learning using multiple intelligences based on student needs and interests. By co-creating a topic-based unit in science, social studies, or their specific subject discipline for subject specialists, and then integrating *diverse* disciplinary lenses to enable learning in *diverse* ways, from *diverse* perspectives, teachers can naturally bring out these multiple intelligences, and allow students a variety of learning options. For example, a topic in science can be enhanced through research and writing, experimentation, mathematical expression, and exploring social implications and applications. For some it may be far more meaningful to draw rather than write notes. Drama, physical movement, art, and music offer diverse ways to explore and express learning. Expanding beyond the classroom walls will enable diverse ways of learning by doing. Traditional Ecological Knowledge and Indigenous ways of teaching and learning will help bring in *diverse* worldviews and place-based learning in decolonizing education.

In transitioning traditional classroom-based teaching to incorporate the active learning cycle into a unit of study, teachers could try out *diverse* teaching methods by taking the students out into the school grounds initially, and progressively into the community and nearby natural areas as confidence in place-based experiential teaching and learning develops.

These *diverse* locations can bring in opportunities for community members to contribute *diverse* cultural perspectives and ways of knowing. As teacher and learner confidence grows, incorporating inquiry-based learning also empowers students to take the lead in developing *diverse* competencies and multiple intelligences.

Cycling and Feedback

Keeping in mind the ecological principles of *cycling* and *feedback* are key to transitioning teaching and learning practices. Building reciprocal relationships and learning communities is dependent on creating learning cycles through constant *feedback* from students, parents, administrators, the community, and professional colleges. It reminds the teachers, as co-learners, that transitioning is an iterative, collaborative, transitional process.

Transitions in teaching and learning can be further supported by assessment practices that are process-oriented to help teachers and learners identify competencies that can be strengthened by revisiting learning opportunities to further develop and strengthen their competencies in different contexts. *Diversity* in *feedback* is important. Self-assessments help empower teachers in developing new teaching approaches and help learners become self-directed, empowered learners, able to set their own learning goals. *Feedback* from teachers and from students to other students helps expand a students' awareness of their learning potential, progress in developing diverse competencies, and further educational opportunities; and *feedback* from parents and students to teachers helps teachers gain a greater appreciation of the effectiveness of their teaching approaches, as well as ideas on how to make learning more interesting and relevant. These reciprocal *feedback* loops are essential in developing teaching and learning as a collaborative partnership that enables transitioning to sustainable education.

Adaptation and Emergence

Adaptation and *emergence* are what transitioning is all about. Being open to what needs to *adapt* in shifting from our traditional ways of teaching and learning; and then having to courage to try something new will lead to further *adaptations* and *emergence* in transitioning to sustainability teaching and learning. To set this process in motion, the active learning cycle can be applied to bring about transformational change.

Starting with direct experience encourages teachers to try new teaching approaches, letting go of traditional teaching methods based in the top-down control teachers are used to, and comfortable with, in determining what students will learn, when, and how. To develop a more collaborative approach, inviting students into the process and trying out teaching a topic-based unit are good first steps. Based on how these first efforts unfold, critical reflection will help teachers step back and evaluate what went well, what didn't, and why, relying on professional learning communities with feedback from students, parents, other teachers, and administrators. These insights, recorded in a learning journal, can lead to further considerations and an abstract understanding of how ecological principles, as elaborated above, can be integrated to help adapt and improve problematic aspects identified in previous iterations of teaching and/or learning. The teacher and students can then apply these insights to improve further through active experimentation, so that sustainable education emerges organically. Through this iterative cycle of direct experience, critical reflection, abstract understanding, and active experimentation to apply the ecological principles, teaching and learning becomes a collaborative transformative process. The key is for teachers and learners to be open to embracing *persistence, adaptiveness, variability, and unpredictability* to enable creativity and innovation in taking advantage of unexpected teachable moments and learning opportunities.

The Whole Is Greater

In considering the whole being greater than the sum of its parts, it's important for teachers transitioning their practice to recognize they are contributing to a fundamental societal transformation and their actions are not only influenced by the system they are part of, but through their transitioning they also influence the greater whole in helping develop a sustainable society. By creating the conditions for students to thrive, students will be able to develop holistically with multiple intelligences using their head, heart, and hands in developing 21st-century competencies, ecological intelligence, and sustainability as a frame of mind.

Getting Started

Being immersed in a traditional system, teachers can often feel limited by a mechanistic curriculum design, the school timetable when the day is broken up by separate subjects and specialists, and the expectation of assessing predetermined learning outcomes in each of the separate subject disciplines as well as various competencies. This is why it is important to take a pro-active collaborative approach to work with administrators, the union, and colleague, as noted in previous chapters, to address these barriers systemically. While these concurrent changes will take time, it's important for teachers, parents, and students to do what they can, within their circles of influence, with the support of their administrators.

To initiate changes in teaching and learning, it can be helpful to start by taking the students outside for a lesson in or within walking distance of the school grounds, whether that is for a scientific exploration, gathering data for mathematical applications, exploring a social studies topic in the community, or simply writing outside. These shorter outdoor learning experiences help students as well as teachers get used to different learning environments, learning how to facilitate inquiry-based learning that helps students stay on task and learn from Nature, as well as determining what materials and clothing are needed to become comfortable learning outside, being open and place-responsive. While outside, the teacher could keep a learning journal to note what is working, what isn't, how different students respond in different situations, as well as potential integrative opportunities with other subjects.

These experiences can then lay the foundation for developing an integrated topic-based unit. Initially an elementary teacher might schedule a few afternoons a week for their topic-based learning, and then as experience and confidence build, the unit can be expanded to incorporate the traditional morning sessions that tend to focus on language arts and mathematics by seeing how the traditional language or mathematics curricula can be taught through the topic-based unit. Once a teacher sees integrative potential, learning becomes less segregated into separate subject disciplines, to enable learning to develop organically through the active learning cycle, with the various subjects shifting from being the central focus to supporting contextually relevant inquiry-based learning based on the Eco-centric Curriculum Design Framework (Figure 6.1). As there are often separate subject specialists working with the students at different times in the day in both elementary and secondary levels of education, collaborating with these other teachers can extend the integrated teaching and learning through these other disciplinary lenses, as well as strengthen professional learning communities of practice. Once teachers open the timetable to enable interdisciplinary systems thinking in teaching and learning, there are more creative opportunities to learn in the outdoors and community, as well as the class, locating learning in the most contextually relevant locations.

To help teachers transition from teacher-centred to self-directed student learning, many practising teachers have set aside time, starting with an hour a week for independent student projects. Often referred to as a Genius Hour (McNair, 2017), it is a time when students are given the freedom to develop learning guided by their interests, knowledge, and curiosity. As it is open-ended, it empowers students to develop self-directed learning competencies and enables creativity and innovation. By giving students the freedom to explore a diversity of topics in a diversity of ways, based on their interests, teachers learn to take on a diversity of new roles in facilitating, guiding, mentoring, and supporting independent learning. Through letting go of control, just for one hour, teachers start to build trust in students' abilities to take responsibility for their learning. Once teachers see the value in self-directed learning, what started as one hour a week can then expand as students develop their self-directed competencies, intrigued and motivated to develop self-directed projects as an integral part of their learning in relation to, or potentially branching off from, the topic-based curriculum. In supporting these self-directed student explorations, the teacher has very important roles in providing space for self-discovery, suggesting new avenues by considering various subject lenses or possibly encouraging them through stages of the active learning cycle to inspire deeper, more holistic learning.

Through this gradual process, learning becomes more organic, and teaching and learning becomes more fluid, *adapting* and *emerging* in a *diversity* of ways as an *interdependent community* of learners. The day may start out in a shared experiential activity and then diversify as different activities evolve based on experiences and interests of students. This is where the teacher learns to be open to facilitate a variety of learning opportunities. While some students might negotiate and follow up on inquiry-based projects and learning activities, either independently or in groups, the teacher might start a workshop to help develop specific competencies that are needed to further student inquiries or take a few minutes to check in with various students. The teacher's role will change and transform as the students get involved in constructing their knowledge, understanding and evaluating their learning.

As the teacher becomes more comfortable with *emergent*, inquiry-based learning, creativity is encouraged with imaginative, open-ended activities involving the arts, hands-on activities, and critical thinking to enable innovative ideas and empowerment. Throughout the learning process students are honoured with respect and a belief in their abilities to learn and take responsibility as capable contributors to their community and the natural world they are part of. With the ecological principles as their guide, teaching and learning becomes a collaborative partnership, an iterative process of developing sustainable education for both teachers and learners.

Practical Applications

The following are a few diverse examples from post-graduate research students – educators who are using ecological principles to transition their teaching practices in a variety of contexts. Christina Balint is a secondary science educator, looking at how she can incorporate ecological principles in her teaching in Qatar. Jenelle Guichon is a senior secondary social studies teacher who decided to let go of her control in planning curriculum and lessons to see what would happen if she invited the students to co-create a unit based on their personal interests. Nancy Wilson has designed her own programme so as to teach the Alberta Ministry of Education Grade 8 curriculum through outdoor education (OE). Emily Lo is a practising elementary

teacher and outdoor educator in BC, teaching on call in diverse schools
where teachers may or may not be used to educating through the outdoors. From one teaching assignment to the next, she doesn't know until
she arrives what instructions a teacher may leave in their absence, so she has
found the ecological principles can be used to both guide her own work and
contextualize Teachers On-Call (TOCs) in the larger school community.
Finally, April Kuramoto presents a powerful example of how an Indigenous
non-formal educator in a First Nations community has improved her practice through the ecological principles.

Learning through Ecological Principles in Secondary Science

By Christina A. Balint, practising Teacher in Qatar, Graduate Certifi-
cate in Environmental Education and Communication, Royal Roads
University, BC

As a high school environmental science teacher, I believe that the
ecological principles will play a vital role in making the systems in any
environment more understandable for students. I realize that the syl-
labus I have created for the environmental science course that I teach
touches on some of the ecological principles. I believe understanding
and then developing the ability to identify these principles will give
the ecological principles what I call the stickiness factor: concepts that
stick in the minds of students and are not easily forgotten.

With the way that the traditional Western education system is com-
partmentalized by subject and topic, it is a huge task to get students
to realize science is not just a class, but everything around them:
every phenomenon, every stimulus, and every response. The ecological principles are not only suited to Nature but can also be applied to
urban settings as well as everyday tasks and patterns. They will be very
helpful to bridge the gap between "class," "world," and "so what?".

Though I see how certain ecological principles suit certain concepts better than others, I also see how each ecological principle is
connected to each concept covered in my course. For example, when
I explain the Coriolis effect to students, the first ecological principle
that comes to mind is "cycles" because hot air near the equator rises,
cools as it travels 30 degrees from the equator, then sinks back to the
equator ready to begin again. This cycle forms a Hadley cell. This
circular motion is a clear *cycle*, but I can use the Coriolis affect to explain many of the other ecological principles. The Coriolis affect also
exemplifies *energy flow* as heat from the equator is circulated to the
north and south. It also provides an example of *interconnectedness*, as

biomes are established, as are plant, animal, and microbial life, on the effects of this cycling and energetic flow. These flora and fauna *adapt* to the biomes established by the Hadley cells allowing *diverse communities to emerge. The whole is greater than the sum of its parts* as each molecule of air is heated and cooled to form these giant currents of air. *Feedback* from the Coriolis effect comes in the form of the north-easterly and south-easterly trade winds and affects not only migratory patterns of birds but flight times for humans as well.

What I enjoy about the ecological principles is that they can be applied to broad concepts while at the same time be fine-tuned to acknowledge some of Nature's subtler phenomena. An example of this is energy flow. We can discuss energy flow as the death of a deer decomposing into the earth and returning its body and its elements back to it. But we can also look at energy flow as the feeling of calm that flows from a tree to a human in close proximity.

My plan is to start the year with an art challenge of designing simple symbols that best represent each of the ecological principles after teaching the meaning of each. We will hang them somewhere visible to refer to throughout the year as this will help to develop an ecologically-learning-systems-thinking-frame-of-mind. At the end of the school year, they may forget what a Hadley cell is, but they will remember the ecological principles that govern one.

Qatar's desert is not necessarily the most conducive location to help students develop a connection with Nature. I am aware that the desert is still Nature, but it is a far greater challenge to demonstrate systems at play in such a biome, especially to a student-body make-up of 90% non-Qatari. I also have many biophilic students who are accustomed to city life. Many of my students have never touched soil or climbed a tree. Many are afraid of bugs and have no idea how a watermelon grows. I have students who have never slept openly under the night sky and have never seen a shooting star, or any stars at all. Last year we made big strides to increase the biodiversity at our school: we installed an aquaponic greenhouse, a fish pond, and planted our first trees on campus. I will use these new greenspaces as often as possible to help establish a connection between my students and the natural world around them. I will start with the basics: getting their hands dirty and getting them comfortable with bugs. For many students, this will be their first opportunity to explore, learn from, and honour non-human life. Though my resources are limited, I hope to help my students develop a connection and reverence for the natural world that will lead to pro-environmental behaviours in adulthood as is often a result of OE programmes (D'Amato & Krasny, 2011, p. 248).

In her 2015 article *Ecological Literacy in Design Education*, Boehnert goes a step further by combining six ecological principles (networks, nested systems, cycles, flows, development, and dynamic balance) with six design concepts (resilience, epistemological awareness, a circular economy, energy literacy, emergence, and the ecological footprint) effectively re-imagining design as usual and thus business as usual practices (p. 3). As I often integrate STEAM activities into the curriculum, this can be used to narrow the gap between natural and human-made systems as is exemplified in biomimicry. This I hope will bring the ecological principles one step further ensuring that students today make ecologically conscientious business and design decisions tomorrow.

Lastly, I see the nine ecological principles as a way for my students to highlight what they have in common with Nature, rather than only how they differ. This can help us begin to undo the outdated ideas of dualism and set forth on a path of existing in harmony with all beings, sustainably.

References

Boehnert, J. (2015). Ecological literacy in design education - A theoretical introduction. *FORMakademisk*, 8. Retrieved from https://www.research gate.net/publication/282470139_Ecological_Literacy_in_Design_ Education_-_A_Theoretical_Introduction

D'Amato, L., & Krasny, M. (2011). Outdoor adventure education: Applying transformative learning theory to understanding instrumental learning and personal growth in environmental education. *The Journal of Environmental Education*, 42(4), 237–254.

Co-Creating in the Classroom

By Jenelle Guichon, practising Teacher, MA in Environmental Education and Communication, Royal Roads University, BC

The research question being explored through action research for my Master of Arts in Environmental Education and Communication was, *What would an effective unit on the socio-ecological impacts of our food choices look like when co-created by students and teachers?* Prior to conducting this research, I had feelings of apprehension, nervousness, and self-doubt stemming from worries about letting go of control and the unfamiliarity of co-creative processes that are counter to top-down, centrally controlled, traditional education systems. However, through engagement with the process and using inquiry and competency-driven instructional methods, the co-creation process resulted in an emergent learning experience for both myself as an educator and the student participants.

To begin, we worked through a process of brainstorming to establish what the students were interested in around the socio-ecological impacts of food choices. We synthesized and defined inquiry questions and topics of exploration (the "what"), as well as preferred methods of instruction, learning, and assessment (the "how") in co-creating our curricular unit. After our initial brainstorming session, we checked in continuously to assess the successes, challenges, and needs of our learning. This process was circular, progressing from original ideas/inquiry, to learning activities, to feedback and reflection of those activities, to further refine our inquiries and ideas. This resulted in an ongoing co-creative process that highlighted all curricular competencies, including course-specific big ideas and content, and an engaging and meaningful learning experience for the students.

Several ecological principles were embedded throughout this process. While the content of our unit, the socio-ecological impacts of food choices, inherently reflected all the ecological principles, the process of co-creation reflected these principles in a way that could bring them into other fields of study and align education practices with self-sustaining, natural systems. For example, *interdependence* was reflected in the ongoing interactions between the various collaborators of this project. These ongoing interactions formed the unit as it exists. This collaborative approach also highlights the ecological principles of *community, diversity, feedback, adaptation, emergence,* and *the whole is greater than the sum of its parts.* The *diverse,* multiple perspectives, ideas, and voices highlighted throughout the co-creation process allowed for an emergent learning experience that certainly would not have been possible without the collaboration of many *diverse* participants. As the students took their inquiry learning out into their *community* and reflected on their learning, *feedback* led to further *adaptation* of the unit as it progressed. By assessing various learning activities and identifying the successes, challenges, and needs, we *adapted* the learning in ways that best benefited the system. Lastly, the process was designed to be cyclical: by engaging in a circular system, students could pursue the *emergent* inquiries as we engaged in the learning process. This opportunity to begin the cycle again kept students engaged and curious in their learning journey.

As a teacher, I was surprised with how this framework resulted in multiple successes, including increased student engagement and agency, authenticity in the learning process, and providing students with an opportunity to explore how they learn best. By having students accountable and actively engaged, they developed an understanding of what learning activities were most successful at facilitating meaningful learning experiences. This empowered students not only to take ownership of their learning but also to understand themselves as powerful global citizens on a larger scale. Additionally, this model of learning addressed common concerns regarding student apathy in the social sciences and allowed students to engage in competency and inquiry-driven learning in a way that satisfied curricular requirements. This model also facilitated systems thinking for students and myself as an educator. Students could connect what we were learning in the classroom with their own lives. This helped break down the barriers students commonly experience in a rigid and industrialized education system that siloes education into distinctly separate entities. Although this was a unit situated in social studies, it also touched upon and could be expanded into English (writing skills), Science (ecology, climate change, future of food/bioengineering), and Food Studies (food systems, culture, and food preparation). Lastly, as an educator, I experienced immense professional growth navigating and facilitating the co-creation experience.

In addition to these successes, there were several challenges experienced throughout this process. As co-creation was a novel experience to many student participants, suggestions were occasionally not possible to pursue. As such, it was a challenge to maintain student "buy in," while managing disappointments. Time also presented a challenge, as planning activities "on the fly" did not allow for some activities that typically require a longer time frame to plan (for example, guest speakers or field trips). Constraints of the larger education system were also a challenge; for example, lack of funding and a rigid timetable presented challenges for organizing out-of-building activities. Lastly, educators' busy and full schedules made it difficult for collaborative planning or co-creation between educators.

These challenges could be mitigated through increased top-down support, including increased funding, flexible timetables, and/or built-in collaboration time for educators. Additionally, a "culture of co-creation" within an educational *community* is essential for the success of this process. As highlighted above, to be an *emergent* learning experience, co-creation is extremely effective through collaboration, but it requires a supportive network between students, teachers, and administration.

Ecological Principles in Practice

By Nancy Wilson, practising Teacher, Master of Education in Sustainability, Creativity and Innovation Cape Breton University

In order for the education system to change, we need to turn our thinking on its head and get rid of the traditional "school." The underlying mechanistic framework which has conceptualized our notion of education needs to "fall down," and to do this, we need to change our teaching practice. This reflection of how I use ecological principles to develop sustainability in my teaching will highlight some of my successes, obstacles, and what needs to change in my teaching practice.

Some of my successes as an educator include my systems thinking approach through teaching to the ecological principles of *interdependence, diversity, community, adaptation, emergence, and the whole is greater than the sum of its parts.* In 2013, I started a programme I call WILD an acronym for Wisdom, Inquiry, Learning, and Doing. By using outdoor experiences to integrate the core curricula into topics, or inquiry projects, my students are learning as a whole person. I had this vision to create a "whole person" school programme after living and teaching in Australia where I taught at an outdoor campus for year nine students. I brought this vision back to Canada as I had a vision of what I wanted and without a doubt I knew I didn't want to teach in a traditional, mechanistic framework. Whole systems thinking and ecological sustainability gave me the basis for envisioning educational change. My WILD programme is an example of successfully using not only *The Whole is Greater Than the Sum of Its Parts*, but a variety of other ecological principles in order to develop sustainability in my teaching, and to break down the walls that hold us back.

Another ecological principle that exemplifies my overt and emergent curricular practice is *Interdependence. Interdependence* is more than integrating the subjects. In my teaching practice I look to Indigenous cultures to truly understand how a living system educational framework through teaching *interdependence* works. The Haudenosaunee's seventh generation principle assumes the decisions we make today should lead to a sustainable world seven generations into the future (Lapointe, 2020). In order to use interdependence to apply systems thinking, I help the students develop a relationship with the land and learn how we are all related. I believe that the ecological principle, *Interdependence*, not only encourages integrating the different subjects, but it is also important for learning (and teaching) how to be responsible citizens on this earth.

I know I am successful in teaching interdependence because of my student's journal reflections and how they relate to the earth. For example, I always invite knowledge keepers or elders to join us around a campfire or on a hike. One year I had a Stoney-Nakoda water keeper come to camp. She taught my students to talk to water and treat her like a friend. This resonated with my students because months later, when they wrote their final reflection on their "WILD year," many students commented on the impact Ariel had on them and they pledged to respect, protect, and nurture water and treat her as a friend. I thought that was pretty amazing and a great success emphasizing systems thinking and sustainable education and slowly breaking down the walls holding us back.

Another successful ecological principle I use in my teaching practice is *Diversity*. My students learn in many diverse environments, including, forests, mountain tops, river valleys, rock climbing walls, ski trails, canoes, and more. Again, the obstacles here are the mechanistic paradigms with the insurance company not allowing us to go outdoor rock climbing anymore, EVER! And, since our insurance company has changed, now there is not an indoor rock wall that will agree to the "master agreement" put out by my school division. In order for positive change to occur to my curricular practice, I need to be more creative and find ways to adapt to the mechanistic paradigms that are still prevalent in my school division. For example, since we are no longer allowed to rock climb, I should try to incorporate these skills on hikes and find rock faces where we can "boulder" and teach the students basic climbing skills that way. Again, despite these obstacles being there, I think I can slowly break down these walls by continuing to use ecological principles such as *Diversity* to develop sustainability curricula in my practice.

Without a doubt, having knowledge keepers come to my class and camps and field days is also an example of the ecological principle, *Community*. Community is an important construct for sustainability, innovation, and creativity, getting feedback on how well we engage in and promote effective community engagement through our human and more-than-human communities is so important to include in our teaching practice and I feel I am quite successful at this; however, an obstacle is the cost to have an elder visit the class, so again, I am "hitting the wall" with liability and red tape. Undeniably, the old mechanistic linear framework of control and not being allowed to take my students to different "off-site" communities or even have the community members visit the class without a vulnerable sector check is becoming more of an obstacle than it was nine years ago, so I need to figure out how to break down this wall.

Finally, *Adaptation* and *Emergence* are another important ecological principle that resonates in my teaching practice. *Persistence, adaptiveness, variability, and unpredictability* are characteristics developed in my teaching practice through the various camps and field studies we do. For example, students learn how to dress for the weather, work in groups, deal with the unpredictability of situations such as river levels, potential wildlife encounters, and trail conditions and the students adapt. For me, the biggest obstacle is making the connection with this "real life" learning through *adaptation* and *emergence,* to the curricula. In summary, even though my programme leads to life-long learning and empowers students to explore the unknown, the mechanistic learning of separate subjects in classrooms is prevalent in the rest of the school and I need change this and "take down these walls" through *adaptation* and *emergence.*

In conclusion, in order to take down the walls in the rest of my school, and in every other school, I need time to develop a programme so another teacher can pick up and teach this change in perspective. What I also need is to sell this type of teaching practice to other teachers, so they understand the importance of teaching creatively and innovatively using ecological principles. Without a doubt, I need parents to be on board with this change of teaching from the mechanistic root metaphor framework to this eco-centric educational paradigm. In short, I need the persuasive language and actions to convince my school division that more teachers should change their teaching practice to teach in a way that empowers students to be creative, innovative, and able to adapt to and develop solutions as we grow, learn, and conditions around us continue to change. After all, teaching in a creative and innovative way enables sustainability which is what is needed to break down the walls of education and allow for change.

References

Lapointe, J. (2020, January 5). *Address environmental racism today for a better tomorrow.* David Suzuki Foundation. Retrieved November 27, 2022, from https://davidsuzuki.org/story/address-environmental-racism-today-for-a-better-tomorrow/

Ecological Principles as a Framework of Outdoor Education for Teachers On Call

By Emily Lo, practising Teacher On Call, MA in Environmental Education and Communication, Royal Roads University, BC.

In transitioning to sustainability, teachers can get children outside and facilitate learning by connecting with the outdoors. As a Teacher on Call (TOC) I hold a unique perspective in the education system. TOCs are expected to step in for classroom teachers and follow classroom routines, schedules, and lessons. In my three years of working as a TOC, I noticed many schools do not facilitate any sort of Outdoor Education (OE). This idea shaped my research project, exploring how TOCs can implement OE. To answer this question, I collected data of lesson plans given by the teacher, any revisions I made to those plans, reflections on lessons, and limitations to implementing OE that day. Data collection occurred daily (as work became available) over the course of four months. In total, this represents 45 days in 11 classes ranging from Kindergarten to Grade 5, in four different schools. Some schools I worked in have embraced OE in that many teachers facilitate learning outside and connect learning to the land, while other schools have not. My research findings show that TOCs can influence classroom teachers' OE practice through adapting given lesson plans to the outdoors, giving feedback to teachers about outdoor activities, and highlighting the accessibility of OE. This, however, did not come without obstacles such as familiarizing myself with each school's environment, building relationships with students, and juggling class routines and schedules to make time for OE throughout the day.

Understanding the lack of resources for TOCs and the unfamiliarity with teaching outdoors, I am continuing this emergent process by creating a website to help TOCs navigate their OE journey through the transition to sustainability. Part of this resource includes using the ecological principles of *interdependence, community, diversity, cycles, feedback, adaptation,* and *emergence* as a framework for integrating OE as a TOC (see Figure 11.1).

Interdependence

In integrating OE as a TOC, there are three examples of interdependence – between the school system and TOCs, between schools and the outdoors, and between TOCs and students. Schools depend

Emergence

TOCs teaching OE can lead to other teachers' curiosity and interest about outdoor learning

Interdependence

Schools, TOCs, students, and the outdoor environment depend on one another for facilitating play and learning

Community

TOCs are part of the teaching community alongside classroom teachers, administration, and greater education system

Adaptation

TOCs adapt learning activities to the school environment, students, and routines

Ecological Principles as a Framework of Outdoor Education for Teachers On-Call

Diversity

Students, schools, and environments are all diverse- no two lessons/ teaching approaches will be the same

Feedback

The more students interact with nature, the more comfortable they will be in learning outdoors

Cycles

Teaching OE may not happen every day but the learning and experience can impact students' personal growth. Kolb's Active Learning Cycle (1984) can be used to structure OE lessons.

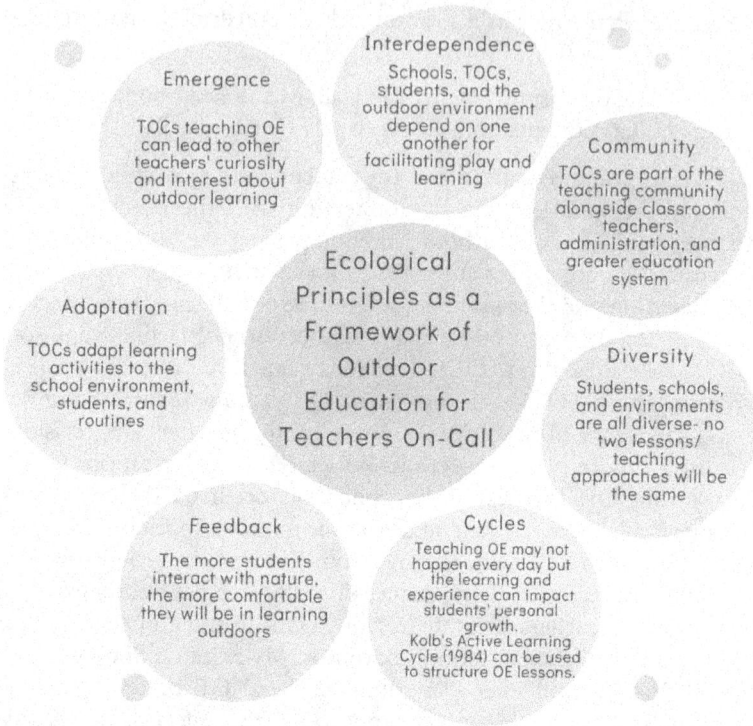

Figure 11.1 Ecological principles as a framework for teaching on call. Created by Emily Lo, included with permission.

on TOCs to fill in for teachers when they are absent and TOCs rely on teachers being away, for work and income. Without TOCs there would be a shortage of teachers and unsupervised children every time a teacher is away. Similarly, there is a relationship between schools and their outdoor environments. Schools depend on the outdoors to provide space for students to learn and play, whereas the space depends on students to maintain it and treat it with respect. TOCs can take students outside to engage in learning with the natural environment while also teaching the importance of sustainability.

In my research, I found that developing effective relationships with students is key to successful teaching on call, especially with the potential of working with new students every day. As a new person entering the classroom community, students may try to push the boundaries by behaving in ways they would not normally behave when their regular teacher is present. Through establishing strong relationships with students, by having open conversations, showing interest in getting to know one another, and outlining expectations early on, I have found students show more respect and value the knowledge you have to share. Connecting with students makes it easier to transition from an indoor lesson to purposeful learning outside, as it impacts their behaviour and actions throughout the day. Maintaining positive relationships between TOCs and students also affects the likelihood of getting asked back to the same class or school.

Community

There are many communities within the education field – schools, teachers, students, and environments. There are also various levels of communities, from individual classes to the school district. OE teachers, as a specialist community, often come together to share resources, knowledge, and experiences, strengthening and supporting one another to continue their practice. Students, teachers, administration staff, and parents are an example of a community within the school, and as they often share something in common, they can come together to create change. TOCs are essential to this larger school community as they need to be familiar with its culture, and educational developments if they are to step in seamlessly, often at a moment's notice.

TOCs are often overlooked in the school community due to the impermanence of our positions. Often, TOCs are not part of the conversations that dictate the way in which schools are conducting their learning. As they do not belong to one school, they are not invited to participate in staff meetings, a space where conversations can spark new ideas of outdoor pedagogy. This, however, does not negate the important work we are doing each day. As a community of TOCs, we can spark change through our work: by sharing it in staffrooms, at staff meetings, and by representing TOCs and OE at district level meetings.

Diversity

TOCs experience diversity in every workday: in schools, lesson plans, students, and the environment. Each of these variables is diverse in their own way, therefore I have come to the conclusion that no two lessons are the same because no two schools, lessons, group of students, and environments are the same. Throughout my research, I experienced diversity in the schools and classes that I taught. Each class was unique, therefore I adapted my teaching strategies to best fit the group of learners. Diversity in learners and environments challenge TOCs to embrace this unpredictability and practice flexibility and adaptability within each teaching day.

Cycles

Each school day can be viewed as a cycle within the larger process: starting when the first bell rings, learning continues each day flowing from one day into the next. The cycles may look different based on changes in routine, activities, and environment but the cycle of daily education is ongoing. Taking learning outside as part of an active learning cycle, by providing direct experiences with natural environments, can impact a student's personal growth, through allowing them to deepen their relationship with place. Students' experiences outdoors, even if it only for one day, can influence their future passions and actions. Becoming the "favourite TOC" for taking them outside is just an added benefit.

Further, through my research I have found that teachers are often hesitant to teach outdoors because they are wary of changing the cycle of already established learning routines. TOCs teaching OE can impact the teacher's education cycle. Realistically, adapting indoor lessons to the outdoors may not occur every day depending on the given day plan. However, if successful, TOCs can influence teachers to integrate it into the daily learning cycle, which can then influence weekly, monthly, and yearly approaches to teaching and learning. Facilitating OE as a TOC may be the nudge classroom teachers need to start implementing their own OE. A change in one teacher's practice can also affect future implementation on larger cycles: as teachers begin to implement it in their own practice, it is possible that the school adopts OE as a regular part of learning.

Feedback

There are two examples of the importance of feedback in the lens of facilitating OE as a TOC. Most obvious is that TOCs provide feedback to the teachers at the end of the day. TOCs give them a glimpse of the activities covered and any challenges or successes. This allows teachers to plan lessons for the future, address any issues with students or parents, and help to shape their teaching practices knowing what works from another perspective. Throughout my research, I gave teachers feedback about how I adapted the lesson outdoors, how the lesson went, and my thoughts regarding future lessons outside. This has helped teachers get ideas for how they can also teach outside in ways they had not thought before. It also gave teachers insight for how to best support their students while facilitating learning outside (such as extra supplies, previous expectations, and established boundaries). Secondly, feedback is apparent in that the more students interact with Nature, the more comfortable they become in learning outdoors. Through experiential activities outdoors, students deepen their relationship with the place they learn and play in every day. As more educators adapt to OE, students will become more familiar and connected with their unique outdoor environment.

Adaptation

Based on a wide diversity of variables, TOCs are tasked to teach to the diversity, adapting teaching approaches, learning goals, and activities to fit the students' needs. Adaptations typically happen on the day of or throughout the teaching day, therefore adapting to new teaching environments can be challenging for a TOC, especially as there are often unknowns when walking into a classroom. As mentioned in the interdependence section, TOCs need to manage behaviours, modify plans to fit learners' needs, and form relationships with students, not to mention reconfigure lesson plans given by the classroom teacher, adapting them for OE.

Shifting a typically indoor lesson outdoors is one example of adaptation. Due to the variability in each teaching day, lessons and activities will look different with each group, and the environment learning takes place in. Especially when teaching OE, the lesson will vary based on the school's natural environment and the students' familiarity with the space. TOCs can practise flexibility throughout their day, something that not all classroom teachers feel they can do given the mechanistic system in which they have structured the learning. During my research I noticed that most teachers who do not facilitate OE tend to break the day into learning blocks for each subject, creating a perception that OE cannot be implemented because it does not fall into one subject or align with planned unit timelines. TOCs can use their flexibility to introduce, or reintroduce, OE to students and potentially inspire other educators to adapt their planning to do the same.

Emergence

TOCs hold a unique position of facilitating learning, where added pressures of assessment and unit planning are not part of our tasks. TOCs can try out activities and stretch students in ways that are different from when the regular teacher is facilitating lessons. Adaptations made by TOCs will change the outcome of students' learning and experience, allowing new and deeper insights to *emerge*. Students may be willing to try new things, such as learning outdoors, since the teacher is also new to the class. In my experience, students are quick to point out the differences between TOC's teaching and their regular teachers. Not only is facilitating OE a good practice in flexibility for the students, but it also shows them that learning, in being open-ended and inquiry based, can look and feel different from the mechanistic ways they may be accustomed to.

As mentioned earlier, TOCs teaching OE can also lead to other teachers' interest in facilitating their own OE with their students. When teachers saw me facilitate outdoor activities during my research, they would often ask what I was doing, if the students were still "learning," and how I adapted the given lesson. As other teachers in the school see how students can be engaged and on task while learning outside, it sparked a conversation about how to facilitate this kind of learning in their school environment. The sharing of activities and resources helps to spread the idea that OE can be taught by anyone in any environment because Nature is all around us. Adapting indoor lessons and teaching them outdoors connects the learning with the local flora and fauna, natural phenomena, and relates it to the students' previous experiences to build on their knowledge. This in turn can bring together a community of learners with other human and non-human beings that share the environment.

Example of Adapting a Lesson Based on the Ecological Principles

Table 11.1 is an example of adapting a primary science lesson from indoors to outdoors. The outdoor lesson gets students engaging with their natural surroundings, using their imagination to gain perspective and understanding of the science concept the classroom teacher planned for.

Interdependence

In the example adapted lesson students are tasked to take on the perspective of an animal seeking shelter for the winter. This shift in

Table 11.1 Adapting an Indoor Elementary Science Lesson to Outdoor Education

Given Indoor Lesson Plan	Adapted Outdoor Lesson Plan
1 Read a story about hibernation (e.g., *Hibernation Station* by Michelle Meadows).	1a Take students outside to a natural area where animals might hibernate (e.g., the back forest).
2 Discuss: Where do animals hibernate? What do they need to hibernate? (Dens are typically hidden, insulated, enclosed, and have a food stored). Write ideas and key words on chart paper.	1b Read a story about hibernation (e.g., *Hibernation Station* by Michelle Meadows).
	2 Look around the environment, ask the students: If you were an animal...where could animals hibernate? What would you need? (Dens are typically hidden, insulated, enclosed, and have a food stored).
3 In journals, have students write 5 sentences and draw a hibernation den. Encourage them to label their drawings and what makes a good den.	3 Have students make their own miniature hibernation den out of natural objects (they can either work in groups or individually to build small dens). Remind them: what is needed for a shelter? What is the proximity to food and water?
4 Invite students to share the hibernation den they created with partners or the class.	4 Walk around as a class and have students tell the class about their dens, what makes it a good den? Invite students to give feedback (e.g., with two stars and a wish).
	5 After seeing other students' dens, have students return to their den and assess whether/what changes can be made to make it even better.
	6 Extension 1: have students draw and write about the den they built.
	7 Extension 2: have students measure (using a unit of your choice) the distance between their den and a food/water source.

Created by Emily Lo, included with permission.

perspective allows students to see the school environment in a different light than they typically would when playing outside for recess. Through using their imagination, natural objects start to have different uses and values. The shift in perspective that is practised through outdoor play helps students understand the interdependency between humans and non-humans.

This outdoor learning also lent itself to subject integration, teaching more than one subject in a lesson. In the example lesson science is the main subject; however, there are aspects of language arts, maths, and art integrated into the lesson and the extension activities. This allowed the students to gain a deeper understanding of concepts as systems, considering wholes and relationships. In my experience, inquiry-based pedagogy often led to further knowledge on the topic, based on concepts from other subjects, due to the question-based process. Integrating various subjects into an OE lesson demonstrates systems thinking and builds students' awareness of the interconnectedness of all things.

Community

Not only is a class a community of learners, but they are also part of the greater community of living and non-living beings. Outside explorations and discoveries, as described in the example lesson, can encourage students to practise their imagination and connection to the place through the experiential activity. Simultaneously building their relationship with the land while gaining knowledge is a win-win.

Diversity

Students' needs and abilities are diverse; therefore, TOCs must juggle this diversity and come up with activities that all students can participate in. Learning often looks different for each student: they can be visual, auditory, reading/writing, and/or kinaesthetic learners. Activities that include various ways of learning are more likely to keep all students engaged. For example, in the above outdoor lesson, it allows visual and auditory learners to engage with content through listening to the story and sharing their dens with others while kinaesthetic learners can learn through building. When adapting lessons, keep students' abilities and comfort levels in mind.

Cycling

In each school day, students cycle through various subjects and activities. Through my research, I have come to find that the process of facilitating OE and the learning content shape the implementation of OE. If the TOC is unfamiliar with student behaviours, outdoor environment, or class routines, it is more challenging to adapt a lesson to take place outside. In order to cycle between different learning environments, it is important for TOCs to arrive early to the school to become familiar with the outdoor environment surrounding the school. Knowing the space in which you are teaching makes it easier to transition from the classroom to outdoor learning then back to the classroom throughout the day.

The learning content and lesson structure also help to ease the transition from classroom to outdoors. The example adapted lesson follows an active learning cycle: starting with a story to allow students to become comfortable outdoors, then using the adaptive learning cycle and inquiry learning to flow through direct experience (building dens), critical reflection (through inquiry-based questions), abstract conceptualization (questions and peer feedback prompting deeper reflection), and active experimentation (redesigning dens after peer feedback).

Feedback

If choosing to shift a lesson outdoors, we can acknowledge that this may be a new experience for students who are not familiar with outdoor learning as part of their everyday learning. On the other hand, students may be very familiar with learning outside and understand how to respectfully interact with Nature. My research findings show that familiarity between students and the natural space shapes the TOCs comfortability and lesson planning when taking a group outside. When adapting indoor lessons for my research, such as the example lesson above, I shifted to inquiry-based learning because it elicited feedback from students through questions, discussions, and observations. This would give me insights to adapt to make students more comfortable or deepen the learning. In my end-of-day note to the teacher, I would share my adaptation of the lesson plan, students' feedback/learning from the activity, and suggest a follow-up activity for when they return.

Adaptation

As mentioned in the diversity section, there are many diverse aspects within each teaching day. Our role as TOCs means we will be adapting many aspects of the lesson to work for the students and teacher. Using part of the given indoor lesson saves the TOC from planning a whole new outdoor-oriented lesson, especially when time is limited. In the above adapted lesson, there is an example of adapting part of the teacher's plan to fit an outdoor learning environment. Typically, books, discussions, and activities can easily be taken outdoors. In my experience, shifting outdoors typically requires extra supplies that would not be necessary if conducting a lesson indoors: a clipboard for a hard writing surface, a shared box of pencils and erasers so that students do not have to bring their whole pencil case, sit pads for students' comfort, and weather appropriate clothing.

Throughout my research period, I also adapted lessons as they unfolded based on student feedback and discoveries. For example, in the above adapted lesson, if students are keen on building hibernation dens, following the class walk-through, they could make changes to their den based on the feedback they received from their peers. If students are not interested in further construction, they might draw or describe their den, highlighting what makes them special. Adaptations made by TOCs will change the outcome of students' learning and experience, helping them stay engaged in their learning. The ongoing teaching adaptations show the accessibility of OE, and that teachers can shape their OE pedagogy in a way that works with their practice.

Emergence

Any kind of structured or semi-structured outdoor play is a great activity for TOCs to facilitate as it requires little planning and allows students to connect to their environment in their own way. Emergent learning happens from these interactions so TOCs need to be open to what arises for the students. Emergence can look like a sharing of knowledge between Nature and students as well as student to student. For example, building mini hibernation dens allows students to forage for materials through exploring and familiarizing themselves with the environment. In my research I found that an educator's responses to students' inquiries and findings can enable further reflection of their learning. When building hibernation dens, students may notice a tree that has lots of pinecones underneath it and it becomes the go-to spot for collecting pinecones, which leads to conversation about where to best locate their den and what makes it a good spot. The open, emergent outdoor experience makes the learning more memorable for students, making them want to learn outside more often as they develop deeper connections with their natural environment.

Ecological Principles for Health Workers in First Nations Communities

By April Kuramoto, MA in Environmental Education and Communication, Royal Roads University, BC

In my professional practice, as a health worker for a beautiful and closely connected Coast Salish First Nation community, which

I do not name here out of respect for their privacy and sovereignty of the community, ecological principles can be utilized to facilitate transformative experiences when working with both individuals and groups. My work is funded to work with families on reserve, and the multi-generational family dynamic often means I work with infants, youth, parents, elders, and everyone living in the community.

When working with individuals, I am often visiting mothers struggling with the stresses of young families. Some of the families have up to nine children under the age of 12 and are dealing with extreme poverty. Ecological principles offer a framework for our walks and meetings, where I could initiate uplifting conversations around the nine principles, once per visit. My intention would be to encourage people to share their experiences, so I could then reflect back their innate gifts, resiliency, and inner resources. Connecting with Nature may improve a personal sense of place within the community, bolster mental health, and offer light and exercise.

I also facilitate healthy family groups, where we have guest teachers, make traditional art and foods facilitated by knowledge keepers, and I offer workshops on topics for healthy living. Ecological principles could be used in a variety of ways, for example, we could spend an entire month exploring energy flows. Below is a list of examples on how ecological principles could inform group sessions, and there could be many more ways as well.

Energy flows: Baby massage, family yoga, and songs highlight how energy flows between children and their caregivers. We could walk through their unceded and traditional territory and look for examples in Nature with our cameras and end with discussing how to ground and interrupt that downward spiral that can contribute to a difficult day.

Interdependence: We could begin with a sharing circle discussing how the day flows in each home, showing how actions affect each other and we are interconnected. For example, if one parent goes to the food bank, and another person makes dinner and the kids do the dishes. I could give handouts on age appropriate chores and invite an elder to talk about the importance of everyone helping out and working together. We could be on the land and complete with discussing how we are all connected and we all win by helping out.

Feedback: This is a great day to go over non-violent communication and how to give safe and constructive feedback. I have a simple handout for families, as well as can encourage them to arrange a visit with me to practise communications skills. This transitions well to discussing tension, where we could then move towards pine needle weaving or making cedar bracelets, which highlight finding balance. Teachings here are to have a good heart and mind, so that energy is interwoven into what we create. The art portion of this workshop is stretched over a few weeks, harvesting materials, preparing them, working with them, and then giving the first offering away as a gift. I encourage each participant to make two, so they can keep one if they choose. Each week we could open with a walk to clear our minds, and close with discussing how Nature is always changing, and ask the children to observe what may be shifting week to week.

The whole is greater than the sum of its parts: This lesson can be worked into our healthy cooking day; we can prepare a soup from the community garden together for our lunch. Each family can harvest, clean, and chop a vegetable and we can cook then eat it together. Leftovers can be sent into the homes. We could go outside and explore the land where this ecological principle can be observed, and discussing the importance of family and how each person is important. Families are like the soup, all the ingredients together collectively make something much more delicious, dynamic, and nourishing than each "ingredient" individually.

Cycling: This is an opportunity to discuss cycles: seasons in Nature, moon phases, stages of life, child development, women's moon cycles, or even cycles in relationships. Each of these topics could be a lesson so the group is always adapting, and each topic could have a handout to guide journalling or self-reflection questions. We could walk through the traditional territory and look for examples of life cycles in Nature to sketch or draw. We can observe where plants are at in any cycle: new growth, full bloom, going to seed, etc. It would be beautiful to complete with an elder offering a prayer in their Indigenous language to support the cycles of our lives, families, and community.

Adaptation: Today we could enjoy a longer walk through traditional territory and look for examples in Nature of plants and animal kin adapting to their surroundings; a longer walk, with an elder to share teachings, healing knowledge, and how people have adapted through history and the challenges brought with colonization. If we have permission to harvest, we could make tea to share and healing salves to take home as gifts.

Emergence: Here I could work in conversations on trauma, presenting on some of the behaviours associated with a trauma response. I could guide clients through a meditation of choosing one manageable trauma to work with, and shaping the pipe cleaner into a symbol, nothing graphic. For example, someone who was abused made a watch because she remembered looking at it as a focal point. We can then discuss strengths and resiliency. We could walk outside to a prayer spot and look for examples of Nature in a state of emergence, like us. Ending with a quiet place to sit, pray, clients are invited to re-shape their pipe cleaners to a symbol of their strength or skill gained. It is important that cultural supports are present for this work, to offer a brushing off, prayer, and support for the group.

Community: The entire interdisciplinary health centre team is committed to fostering and supporting a healthy community. We could host a community meal, where the youth are asked to serve the elders first. We could serve traditional foods as medicine and harvested tea to share to complete our evening around the fire, with stories from elders and anyone willing to share songs.

Diversity: Discussing how diversity strengthens ecological systems and their ability to survive, and how that applies to community, we could end this series with a few weeks to make and paint drums, admiring how we share the same materials but each drum is so uniquely beautiful looking and sounding. Alive with their own gifts, just like us. We could look for examples of diversity in Nature and have a song circle to close.

This knowledge has always existed in the teachings and communities I work within. The Ecological Principles Framework acts as a guide to assist me in facilitating these opportunities as a health worker with the honour and professional role of serving Indigenous families and communities. Huychq'a, April Kuramoto.

Conclusion

These suggestions and examples educators have developed to help transition teaching and learning based on the ecological principles are offered to help inspire educators – both formal and non-formal – to take their first steps in transitioning their practice, with diverse cultures at elementary, secondary, teacher training, and community levels, towards sustainability education. This is a continual process of *adaptation* and *emergence* as we gain experience, insight, and conditions around us change. Using the principles of sustainable living systems provides the necessary framework to guide and navigate this active learning cycle.

As noted in the Introduction to this book, numerous conditions around education are changing, challenging how we educate. As conditions around us change, we need to adapt and emerge new solutions based on sustainability. As a recent example, as this book goes to print, artificial intelligence (AI) is challenging how we teach and learn. AI is being used to drive innovations in engineering and medicine but there are also concerns as it is challenging our traditional conceptualizations of teaching and learning. Some teachers are using AI to help write lesson plans and students are using AI platforms such as ChatGPT to not only research but also write essays or to get what is perceived to be the "right" answers. As a result, educators are wondering how to respond. Staff meetings are showing opinions ranging from a traditional mechanistic perspective that tries to maintain the status quo, trying to control and stop its use by planning and imposing new rules of engagement with the technology, to more open perspectives that see it as a tool to be used, adapting teaching practices to focus more on developing competencies. Rather than using dualistic "either/or" thinking, it's important to use "yes/and" systemic thinking. As AI is now part of society, we need to have broader conversations that will contextualize it and bring in diverse ways of knowing. With this latest change in the conditions that surround us, we can once again use the ecological principles to guide us. In developing sustainable education, how can AI be used effectively?

Using an eco-centric perspective, AI can be seen as a powerful tool when grounded in an eco-centric context to augment teaching and learning. Relevant output depends on the quality of the input; therefore, students and teachers need to develop ecological literacy and their 21st-century competencies to use it effectively. The user's critical thinking, media literacy, and ecological intelligence are needed to discern how effective results are for learning and progressing sustainability. As it relies on and "learns" with input from a diversity of perspectives and background knowledge, it models and reinforces interdependence, and learning through community collaboration. This expands learning beyond classroom walls and enables pluriversalism. Putting ecological principles and the need to develop sustainably into the algorithms can ground AI in helping further sustainable education.

AI is challenging us to adapt and emerge new teaching methods as the societal conditions change. It does not replace, but highlights, the importance of learning through direct experience with our human and more-than-human communities. Using AI can potentially add to subsequent stages of the active learning cycle by extending critical reflections, abstract understanding, and ideas for active experimentation: a tool used to extend experiential, place-based, inquiry learning. Incorporating AI into inquiry-based learning by contributing to teacher and student insights, using AI can empower learners to be part of larger learning communities and self-direct many of their inquiries. Further discussions and feedback on the relevance of AI's output with colleagues and fellow learners are essential in developing competencies to deepen understanding and determine relevance through critical thinking, and an iterative cycle of *feedback*, and *adaptation*. Grounded in the context of sustainability and what is needed to transition to sustainable practices, using AI as an *adaptive*, collaborative teaching/ learning tool can open new options that may lead to the *emergence* of innovative ideas. However, the use of AI as a teaching tool also highlights the primary importance of grounding learning in the ecological principles and experiential connections in the natural environment in developing ecological intelligence and sustainability as a frame of mind, necessary foundations to discern the relevance, and value of the output. With this eco-centric foundation, educators can guide our use of AI as a useful tool in transitioning to a sustainable society.

As we've seen throughout this chapter, the ecological principles provide a powerful framework and foundation for teachers and learners to navigate both mandated subject-centred curricula and 21st-century competencies, as well as societal innovations and community-based needs. Initiating transformative practices at this smaller scale in a classroom, school, and community has the power to influence others across and between levels in the educational panarchy in developing sustainable education.

References

McNair, A. (2017). *Genius hour: Passion projects that ignite innovation and student inquiry.* Prufrock Press Inc.

Sterling, S. (2001). *Sustainable education: Re-visioning learning and change.* Green Books.

12 The Whole Is Greater than the Sum of Its Parts

In education we have been tinkering around the edges over the past 50 years of my educational career, akin to rearranging deckchairs on the titanic, or as David Orr (2003) has said, we've been trying to walk north on a southbound train. Simply greening school grounds or trying to implement innovative pedagogy in a mechanistic system is not enough. We need to recognize the problematic mechanistic root metaphors that are undermining innovation and maintaining the status quo. As noted in Chapter 1, and shown throughout the book, it's time we build a new ship that can help us navigate the complexity and rapidly changing currents in the 21st century. An eco-centric paradigm supported by eco-centric root metaphors is capable of supporting and charting this new course in education and society.

Developing sustainable education is a holistic, iterative, collaborative, adaptive, emergent process. As we've seen throughout this book it is a process of re-thinking all aspects of our educational system in working collaboratively with all multi-stakeholders: educational researchers, government officials, administrators, community members, unions, maintenance departments, teachers, students, teacher trainees, and their instructors in transitioning policy, administration, and leadership; buildings, grounds, and resources; curriculum; as well as teaching and learning. As Biggs et al. (2015, p. 259) recognized,

> As SES [social-ecological systems] are highly interconnected systems, the properties and processes associated with the different principles do not become effective in isolation from each other. Applying any one principle in isolation will rarely lead to enhanced resilience of ecosystem services. For instance, polycentric governance and effective learning both depend on the social capital and trust developed through participation, whereas connectivity may not enhance resilience in the absence of diversity among nodes.

This shows how essential it is to incorporate all ecological principles to enable systemic change.

DOI: 10.4324/9781003389590-16

In the process of emergence to sustainable education, it's important to recognize and give credit to all those teachers, students, administrators, and supportive organizations at all levels of education as well as all the curriculum theorists and researchers who have each advanced education for sustainability in their own ways, often in silos. Unfortunately, many of these initiatives have been limited or subverted by the mechanistic educational system, yet their examples still remain and resonate, offering ideas that can succeed in a supportive culture. In some cases where there has been support, we've witnessed some systemic changes at lower levels in the educational panarchy. The examples included in various sections of this book are just a few examples of the many brilliant initiatives that have already been and are currently taking place.

This book is intended to help build on this valuable work by providing a unified framework that can pull together and strengthen these pioneering initiatives with support across the educational panarchy, helping create systemic change in motivating and providing guidance for all to be part of this great transitioning. With the wisdom imbedded in the ecological principles of sustainable living systems, we have a coherent, holistic framework needed to guide and support transitioning from mechanistic industrial education to eco-centric educational systems that support rather than subvert innovation, as we adapt and emerge into the 21st century with the increasing need to develop a sustainable society. Based in an eco-centric paradigm, the ecological principles of *interdependence, community, diversity, energy flows, cycling, feedback, adaptation, emergence* and *the whole is greater than the sum of its parts* provide the framework to guide this transformation to sustainable education.

This whole system approach is essential as all aspects of our educational systems and society are *interdependent,* forming a *diverse* learning *community* needed to provide relevant *feedback,* in *adapting* and *emerging* sustainable education that can support transitioning to a sustainable society where all can thrive. When all levels in the educational panarchy work together in creating transformative change, there is synergy to support transitioning at and across all levels in the system so we can address the needs of educating in the 21st century, in the context of society's imperative to develop sustainably. An exciting aspect of this systemic change is all those involved in education influence and are influenced by scales above and below in the panarchy, such that change can be initiated from any level, and initiatives become supported by the larger system.

Discussions and professional development are needed at all levels so everyone understands why our industrial schooling system, designed for the industrial era, is outdated and problematic, undermining the change that is needed in developing a sustainable society; and how they can be part of educational transformation. This professional development shines light on how all can transition their practices and initiate changes that can influence others across and between levels in the educational panarchy. The ecological principles of sustainable living systems provide a consistent, simple, elegant, eco-centric framework needed to guide educational transformation in all sectors from primary to secondary to post-secondary education, both formal and non-formal. Nature has developed the principles that lead to sustainable, resilient systems, able to adapt as conditions change. In mimicking Nature, by applying the ecological principles of sustainable living systems, we have an opportunity to develop the professional insights and expertise to transition education's organizational structure, administration, leadership, and policy; our buildings, grounds, and resources; how we organize curriculum; as well as how we teach and learn, in developing a sustainable educational system to support change such that the system learns, adapts, and emerges as conditions around us are changing.

As we've seen in Parts III and IV the ecological principles inform how all sectors in our educational system can support transitioning to an educational system based on an eco-centric paradigm. Administrators have a key role to play in transitioning from traditional centralized controls based on plan-and-impose administration, to enabling collaboration and change through an organizational structure based on complex adaptive systems thinking, thereby creating a dynamic, interactive, educational panarchy that can *adapt* and *emerge* in developing sustainable education. Positive leadership, servant leadership, authentic leadership, and collaborative leadership are innovative approaches to bring in all multi-stakeholders into the adaptive process and support learning so all can thrive. Also key to supporting change is for administrators to work with government officials in revising policies to align with and support education transitioning to model sustainable living systems.

In developing effective curricula, it's essential to address both the hidden and overt curriculum. In creating supportive contexts for sustainable education, where learning happens - the school buildings, grounds, resources, and community need to aligning the hidden curriculum with the overt eco-centric curriculum, so as to support rather than subvert developing sustainability as a frame of mind. The Eco-Centric Curriculum Framework, detailed in Chapter 6, has been offered to guide teachers in developing contextually based, decentralized, eco-centric curricula based on ecological principles.

In bringing this interdisciplinary, systemic curriculum to life, teachers, students, parents, and community members are supported by the ecological principles to collaborate in developing 21st-century competencies and sustainability as a frame of mind through adaptive learning cycles grounded in inquiry-based, experiential teaching and learning. This iterative cycle of learning applies to how learning unfolds as well as how teachers, parents, and students transitioning to learning new teaching methods, and ways to learning.

As all aspects of the educational system are interdependent, transitioning in each of these educational sectors and levels, based on the guiding framework of ecological principles, will initiate and support changes across the panarchy. Working collaboratively together through healthy *interdependent* relationships in *community*, enhancing *diversity*, enabling *feedback* across and between all sectors and levels in the panarchy, and engaging in iterative cycles will enable *adaptation* and *emergence* such that the whole educational transformation to sustainable education becomes *more than the sum of its parts*. Although each ecological principle guides transformative change in different ways, they are complimentary and mutually supportive. Together, when applied in all educational sectors, across all levels of the educational panarchy, they enable a systemic response that leads to sustainable education.

Nature has developed sustainable living systems based on the ecological principles in order to thrive. By aligning our educational systems with Nature as our model, mentor, and measure in following the advice of Benyus (2014), we not only develop healthier socio-ecological relationships in schools and society, but we also develop sustainability as a way of thinking, so that through sustainable education, all can thrive and develop the means to participate according to their diverse interests and talents in developing sustainable societies. All that's needed is the courage to take our first collaborative steps.

> Twenty years from now, you will be more disappointed by the things that you didn't do than by the ones you did do. So throw off the bowlines. Sail away from the safe harbour. Catch the trade winds in our sails. Explore. Dream. Discover.
>
> Mark Twain

References

Benyus, J. (2014). Biomimicry in action. *TED Talk*. https://www.ted.com/talks/janine_benyus_biomimicry_in_action?language=en

Biggs, R., Schluter, M., & Schoon, M. (2015). *Principles for building resilience: Sustaining ecosystem services in social-ecological systems*. Cambridge University Press.

Orr, D. (2003). Walking North on a South-bound train. *Conservation Biology*, *17*(2), 348–351.

Index

For Product Safety Concerns and Information please contact our EU
representative GPSR@taylorandfrancis.com
Taylor & Francis Verlag GmbH, Kaufingerstraße 24, 80331 München, Germany

9 781032 485478